Unseen

Some other titles in the Bloomsbury Sigma series:

Spirals in Time by Helen Scales
Suspicious Minds by Rob Brotherton
Science and the City by Laurie Winkless
Built on Bones by Brenna Hassett
The Planet Factory by Elizabeth Tasker
Catching Stardust by Natalie Starkey
Nodding Off by Alice Gregory
Turned On by Kate Devlin
Borrowed Time by Sue Armstrong
Clearing the Air by Tim Smedley
Kindred by Rebecca Wragg Sykes
First Light by Emma Chapman
Models of the Mind by Grace Lindsay
The Brilliant Abyss by Helen Scales
Overloaded by Ginny Smith
Handmade by Anna Ploszajski
Beasts Before Us by Elsa Panciroli
Our Biggest Experiment by Alice Bell
Superspy Science by Kathryn Harkup
Warming Up by Madeleine Orr
The Long History of the Future by Nicole Kobie
Into the Groove by Jonathan Scott
The Power of Nuclear by Marco Visscher
Please Find Attached by Laura Mucha
Six Minutes to Winter by Mark Lynas
V is for Venom by Kathryn Karkup
Sink or Swim by Susannah Fischer
Ghosted by Alice Vernon
Spinosaur Tales by David Hone and Mark P. Witton

UNSEEN

Blind Spots and Why We Miss What Matters Most

David Lewis and
Keelan Leyser

BLOOMSBURY SIGMA
LONDON · OXFORD · NEW YORK · NEW DELHI · SYDNEY

BLOOMSBURY SIGMA
Bloomsbury Publishing Plc
50 Bedford Square, London, WC1B 3DP, UK
Bloomsbury Publishing Ireland Limited,
29 Earlsfort Terrace, Dublin 2, D02 AY28, Ireland

BLOOMSBURY, BLOOMSBURY SIGMA and the Bloomsbury Sigma logo are trademarks of Bloomsbury Publishing Plc

First published in the United Kingdom 2025

Copyright © David Lewis and Keelan Leyser, 2025

David Lewis and Keelan Leyser have asserted their rights under the Copyright, Designs and Patents Act, 1988, to be identified as the Authors of this work

For legal purposes the Acknowledgements on p. 273 constitute an extension of this copyright page

All rights reserved. No part of this publication may be: i) reproduced or transmitted in any form, electronic or mechanical, including photocopying, recording or by means of any information storage or retrieval system without prior permission in writing from the publishers; or ii) used or reproduced in any way for the training, development or operation of artificial intelligence (AI) technologies, including generative AI technologies. The rights holders expressly reserve this publication from the text and data mining exception as per Article 4(3) of the Digital Single Market Directive (EU) 2019/790

Bloomsbury Publishing Plc does not have any control over, or responsibility for, any third-party websites referred to or in this book. All internet addresses given in this book were correct at the time of going to press. The author and publisher regret any inconvenience caused if addresses have changed or sites have ceased to exist, but can accept no responsibility for any such changes

A catalogue record for this book is available from the British Library

Library of Congress Cataloguing-in-Publication data has been applied for

ISBN: HB: 9-781-39942-238-3; TPB: 978-1-3994-2239-0;
eBook: 9-781-39942-236-9

2 4 6 8 10 9 7 5 3 1

Typeset in Bembo Std by Deanta Global Publishing Services, Chennai, India
Printed and bound in Great Britain by Clays Ltd, Elcograf S.p.A.

To find out more about our authors and books visit www.bloomsbury.com and sign up for our newsletters

For product safety related questions contact productsafety@bloomsbury.com

In loving memory of Dr David Lewis (1942-2025), co-author, colleague and dear friend.

And to Steven and Angela for their unwavering support and practical assistance in writing this book.

Contents

Introduction	9
Chapter 1 – Looking Without Seeing	17
Chapter 2 – Now You See It – Now You Don't	35
Chapter 3 – It's Your Choice – Or Is It?	53
Chapter 4 – Expectation Blind Spots	67
Chapter 5 – You Could Have Fooled Me!	87
Chapter 6 – Blind Spots and Your Alien Brain	123
Chapter 7 – Do You See What I See?	139
Chapter 8 – Why We Never Forget a Face	159
Chapter 9 – Buying Blind Spots	185
Chapter 10 – Business Blind Spots	203
Chapter 11 – Magic Blind Spots	225
Chapter 12 – Banishing Your Blind Spots	251
Acknowledgements	273
Notes and References	274
Index	319

Introduction

Looking without seeing is the most common cause of accidents in the home, on the roads, at sea and in the air. Our failure to see what matters most influences our ability to make friends, fall in love and pursue successful careers. It also influences what we buy and how we vote. These factors can even shape our social attitudes, political priorities and religious beliefs.

As technology becomes increasingly involved in our lives, opportunities for others to exploit our blind spots also increase significantly. By using tried-and-tested techniques, many of which are also used in magic and illusions, such as distraction, misinformation and playing on our expectations, others find it even easier to persuade us to see only what *they* want us to see while ensuring everything else remains unseen. This book explains the psychology behind such baffling techniques and illusions, which are illustrated with photographs and short films that can be accessed via the URLs provided or by scanning the accompanying QR codes.

Blind spots, in both magic and real life, distort our view of the world in one of two ways: by preventing us from seeing what is there or by encouraging us to see what is not. Throughout history, magicians and psychologists have collaborated to unravel the mysteries of the human mind. In the 19th century, for example, the French psychologist Alfred Binet, inventor of the first IQ test, whose pioneering studies we describe in the first chapter, partnered with magicians to uncover the psychological principles behind classic magic tricks and illusions. Magicians are experts at persuading audiences to see things that never happened and to fail to notice what happens right under their eyes.

During a rebellion in French-controlled Algeria in the 19th century, the French government sought the assistance of magician Jean-Eugène Robert-Houdin. The father of modern magic, he used a blend of psychology and forward-looking technology to undermine the rebel leaders' power and authority by using powerful electromagnets to make it appear that the Marabou chieftains had been robbed of their strength when unable to lift a delicate wooden chest.

Similarly, during the Second World War, British military intelligence employed magician Jasper Maskelyne to help devise various forms of misdirection, such as ruses, deception and camouflage. Among many other illusions, he and his team of fellow magicians drew the enemy's attention to a reproduction of Alexandria harbour on the north coast of Egypt. A few miles away from the real harbour, on Maryut Bay, they constructed fake docks and ships out of cardboard and canvas. Shadows, ground lights and explosive charges were used to mimic an under-siege location when viewed by enemy pilots. Maskelyne's team's illusion attracted enemy pilots' fire over several nights.

During the 1960s, the great American magician John Mulholland was involved with the CIA's top-secret Project MKULTRA, a human experimentation programme designed to train agents, or 'operators', in often abusive and illegal methods of interrogation.[1]

While magic's relationship with psychology has evolved considerably since those days, one thing remains clear: magic offers fascinating insights into the intricacies of human cognition and behaviour. The techniques and principles used to create illusions and manipulate perception illustrate the speed and ease with which the most observant and intelligent spectators can be deceived.

Magicians exploit the brain's natural tendency to recognise and follow patterns and automatically filter out anything it deems unimportant. They hack into our brains at a precise moment for each trick or illusion. That is fundamentally how all magic works. By understanding and leveraging our cognitive processes, magicians guide the audience's attention away from the secret workings of the trick, making the impossible seem possible.

Just as the techniques of the magician – such as misdirection, forcing and priming, all of which we will discuss later – can be used to shape an audience's perceptions and decisions, the principles behind those same techniques can also be applied in everyday life to influence how we think, feel and act. As a result of understanding the psychological mechanisms behind magical techniques, people can acquire a more critical and discerning outlook on information they encounter daily.

Understanding the concept of misdirection helps people identify when their attention is being deliberately diverted away from important issues or when crucial information is being obscured. Awareness of priming effects can make individuals more mindful of how seemingly unrelated stimuli can subtly influence their choices and preferences. Recognising the power of social cues can help people distinguish between genuine and manipulated social interactions and make more informed decisions based on their judgement rather than external influences.

As we bridge the gap between the art of magic and cognitive psychology, we can illuminate our blind spots and unlock our true potential by seeing beyond our limitations. With this unique insight, we gain the power to confront the hidden influences that shape our perceptions and decisions, enabling us to break free from the illusions that have held us back. We can open our eyes to the truth that has always existed. With the tools of magic and psychology at our fingertips, we can finally see what was previously unseen.

About the authors

David Lewis: I have long been interested in how we see the world. As a young man, I dropped out of medical school to study photography and for some years worked as a photojournalist in Fleet Street. I covered assignments in many parts of the world, including war zones, with my photographs appearing in major UK, US and European magazines, such as *LIFE*, *Paris Match*, *Stern* and *Oggi*.

In my early thirties, I returned to the academic world to study clinical psychology. During my second year of study, I became aware of and interested in blind spots after a somewhat embarrassing experience. One shared with a dozen other students.

We were attending a lecture on *diabetes mellitus*. As the Latin name suggests (the phrase translates as 'the passage of sweet urine'), the condition is characterised by the honey-like smell and taste of the sufferer's urine, which, before modern tests, doctors could only detect by inserting a finger into the urine and tasting the results. Hence the somewhat derogatory 18th-century name they acquired: 'piss doctors'.

The professor instructed us to bring a beaker of our urine to the lecture hall. Having explained the archaic testing procedure, she stuck a finger into her urine-filled beaker, put it in her mouth and sucked. She then told us to do the same. Our expressions contorted into varying degrees of disdain and disgust as we all did as we were told.

Only then did she reveal that her purpose was to illustrate the importance of observation in medicine. Had we paid attention, we would have seen she had inserted her index finger into the urine and the middle finger into her mouth! I later learned that medical lecturers had employed this deception since the 1930s. But if not original, it was highly effective and must have fooled thousands of students over the years.

This demonstration sparked my fascination with the idea that we often need help to see what matters most. Driven to understand the factors influencing our perception and decision-making, I conducted postgraduate research in the Department of Experimental Psychology at the University of

Sussex. I later set up the research consultancy Mindlab International in the university's Science Park.

Keelan Leyser: My introduction to magic came through watching Las Vegas magician Lance Burton on the TV show *Knight Rider* when I was a young child. I received a magic set after begging my parents and soon discovered I had a natural ability and love for performing tricks. This childhood hobby became an obsession as I began to understand how magicians could make audiences see exactly what they wanted them to see – and miss what they didn't.

I was fascinated not just by the tricks but also by how a skilled magician could seamlessly control an audience's perception without their awareness. This interest drove me to investigate how our brains can be so easily fooled.

Over decades of mastering the skills of a magician, mentalist and pickpocket, I've observed how deception operates through predictable patterns of human attention and expectation. Misdirection isn't simply about distraction – it's about exploiting specific cognitive limitations consistent across audiences and cultures. Whether secretly removing a watch from someone's wrist or making a pack of cards appear to turn into a block of glass in someone's hands, the principles remain the same: the brain's attention can be redirected through calculated psychological triggers.

Through my extensive performing experience, I figured out how to exploit these psychological blind spots by directing attention. A quick look here, a small gesture there – and suddenly, the audience is looking exactly where I want them to. I can guide their focus away from the actual workings of the trick, which happens elsewhere, unnoticed. As a pickpocket, I've used these principles of misdirection and sleight of hand to remove people's watches or wallets without them noticing or feeling a thing. People tend to have a narrow focus, often failing to notice changes or actions outside their immediate attention. I can divert their attention from what is happening around them by creating distractions and controlling their focus. It's all about walking that fine line between what people notice and what slips right past them, guiding people's attention exactly where you want it, while keeping the real work entirely out of sight. When executed perfectly, these techniques create an illusion of impossibility, leaving people amazed and questioning their senses.

Psychology, in particular, benefits from the reliability of these effects. Unlike many psychological phenomena that prove difficult to replicate in laboratory settings, magic tricks consistently produce the same perceptual blind spots, even when audiences know they're being deceived. Magic tricks expose how the human mind works in ways that most scientific experiments can't match. No matter how often I perform a trick or how aware my audience is of being deceived, they still fall for it night after night. This resilience is something that scientists would struggle to achieve in the controlled environment of a lab.

This book shares the secrets behind magic's most valuable psychological tools. By revealing these insights, we aim to make you more aware of your cognitive blind spots and help you develop strategies for overcoming them.

Magic and psychology offer a remarkable way to understand how our minds work. By exploring our cognitive blind spots, we can learn to recognise the subtle ways our perceptions can trick us. These observations help us understand why we sometimes misinterpret situations or make unexpected decisions.

The goal isn't to criticise how we think, but to become more aware of the hidden patterns that influence our understanding of the world. As we learn to spot these patterns, we become better at questioning our initial assumptions and gain a clearer understanding of the situation.

The insights from magic and psychology aren't about finding some grand, mystical truth. They're practical tools for understanding ourselves better, ways to pause and ask, 'Am I really seeing this accurately?' This self-awareness can help us

INTRODUCTION 15

make more thoughtful choices and avoid falling into mental traps we might not even realise exist.

Keep in mind that if professional magicians can so easily and covertly control what people see, or think they see, and how they subsequently behave, then surely anyone could do the same – not to entertain but to exploit. As we demonstrate in this book, they can and do.

Figure 1: David and Keelan rehearse a magical demonstration.

CHAPTER ONE
Looking Without Seeing

Shortly after two o'clock, on a bitterly cold morning in late January 1995, Boston police officer Kenny Conley receives an urgent call. A fellow officer has been shot. Four suspects are fleeing. Spotting one, Conley gives chase. By that time, dozens of other officers have arrived to render assistance. Among them is African-American detective Michael Cox. The other officers, mistaking him for one of the suspects, savagely beat their unfortunate colleague.

'I was hit in the back of my head, and when I turned around, I got hit in the front of my head,' said Cox. 'I kind of remember trying to get up at some point and seeing a silhouette of a police officer.'[1]

Conley admits running past the assault but insists he never saw what was happening. Investigators, prosecutors and jurors refused to believe him. Because he *could* have seen the beating, he *must* have seen it, and had to be lying to protect his fellow officers. Convicted of perjury and obstruction of justice, he received a three-year jail term.[2]

But had he witnessed the assault, or could something have rendered his fellow officer's beating invisible to him?

To answer this question, psychologists Chris Chabris and Daniel Simons attached body cameras to students and instructed them to chase after a jogger. They were told to follow him at a fixed distance and count the number of times he touched his hat.

The reason for this strange request was to ensure they focused their attention on the jogger, just as Officer Conley's attention had been on the suspect to ensure he didn't pull out a gun or throw away evidence. About a minute into the run, and slightly off to one side, Chabris and Simons staged a mock fight. 'We had two students beating up a third, punching him and kicking him and throwing him to the ground, explains Chris Chabris. 'In the light of day, 40 per cent still didn't notice the student being beaten.'[3]

Their findings strongly suggested the unfortunate police officer was speaking the truth when he told the court he had failed to see anything. 'I don't know why I didn't,' he told the jury plaintively. 'I wish I had.'

As Ulric Neisser, known as the 'father of cognitive psychology', has pointed out, 'There is always more to see than anyone sees and more to know than anyone knows. Why don't we see it?'[4]

We have all had similar experiences when we have made mistakes, although hopefully not as serious as that of the unfortunate Kenny Conley, by failing to see what mattered most.

Searching a crowded restaurant for a free table, you spot one and swiftly claim it. The following day, friends demand to know why, despite waving to attract your attention, you'd walked past them without a glance or greeting. It had not been discourtesy that led you to ignore them but their temporary invisibility. By focusing your attention on that one free table, you had fallen victim to a blind spot. You were looking at them without seeing them.

As we will explain in this book, blind spots come in many guises. There are occasions, as in the restaurant scenario described above, when we fail to see what's right in front of our eyes. Others arise whereby we remain unaware of a noticeable change, or fail to recall a choice we just made. Or there are cases where we see what does not exist because we *expect* to see it or are fooled by an optical illusion.

In September 2013, a 20-year-old student, Justin Valdez, was shot in the head as he stepped off the San Francisco light rail train. Before the shooting, the suspected killer, 30-year-old Nikhom Thephakaysone, sitting opposite him, drew his .45-calibre pistol, pointed it across the aisle, and replaced it in his pocket on several occasions, at one point wiping his nose with the hand holding the gun. Bystanders who could have prevented the shooting were all so engrossed in their phones or tablets that none noticed the weapon until it was too late.[5]

People have limited awareness of the world around them. They can fail to notice important things and events – an obstacle when walking, other vehicles or pedestrians when driving, warning signals in complex work situations, crimes when focused on different events, or a man with a handgun while riding on a train. These failures of awareness often represent instances of inattentional blindness.[6]

It is easy to dismiss blind spots as rare exceptions to our usually perfect powers of observation. To believe we usually see everything we want and rarely miss seeing anything we need to see. Nothing could be further from the truth. Not only are blind spots far more frequent than we realise, but as the world becomes increasingly fast-paced and technologically dependent, they pose an ever-greater threat to us all: to individuals, companies and societies. With consequences ranging from the catastrophic to the embarrassing.

At home, a frosty atmosphere develops after you fail to notice your partner's new outfit or hairstyle. At work, a colleague is upset when you fail to admire their new car or latest mobile phone. Far more seriously, blind spots cause seven out of ten motoring accidents, some of which prove fatal. They have been held responsible for significant collisions at sea and fatal disasters in the air. They have led to significant errors by those in charge of complex machinery, running nuclear plants, treating medical conditions, giving eyewitness evidence in a court of law, making financial investments, going shopping or deciding which way to vote.

Until recently, a significant proportion of mistakes, some of which ended tragically, were due to sloppy workmanship, poor design, or a combination of both. Today, although technological glitches and computer gremlins still occur, it is increasingly common for mistakes to result from human error. 'Regardless of the activity or task being conducted,' says Dr Graham Edkins, chief executive officer at Leading Edge Safety Systems, 'humans make between 3–6 errors per hour. One study found that aviation maintenance engineers made an average of 50 observable errors per work shift.'[7]

A chilling example of the consequences of such errors occurred in January 2024, when a 'plug door', used to fill in the space where a regular exit door would usually be fitted, fell from an aircraft at 5,000m (16,000ft), creating a hole in the fuselage 'as wide as a refrigerator'. Maintenance engineers responsible for this potentially catastrophic accident had failed to notice four vital securing bolts were absent.

Air incident investigators blamed human error for the near-fatal accident. According to Ed Pierson, a former Boeing senior manager, this occurred due to inattentional blind spots caused by having to meet unrealistic deadlines. 'The corporation is pressuring the factories to produce these planes and pump them out the door,' he warned.[8]

While it is easy to understand why we fail to see something when our attention is focused elsewhere, more is needed to explain why so much of what we look at remains unseen. When, in the early years of the last century, scientists started to take a serious interest in blind spots, they directed their attention to unlikely subjects, such as professional magicians.

Masters of Misdirection

Until the end of the 19th century, most academic scientists considered magic so trivial as to be unworthy of serious investigation or consideration. This dismissive attitude began to change as members of the fledgling profession of psychology realised it might help them uncover some of the mind's secrets. Magic does, after all, involve manipulating the audience's expectations, misdirecting their attention and covertly influencing their decision-making.

Two of the most prominent psychologists undertaking these studies were the French psychologist Alfred Binet, co-inventor of the IQ test, and American Joseph Jastrow, prolific writer of self-help books.

Born in Nice on 8 July 1857, Binet had given up a successful career in the law to devote himself to medical studies at Paris's prestigious Salpêtrière Hospital. In 1891, he began working as a researcher in the Laboratory of Experimental Psychology, becoming its director ten years later.

In 1893, he invited France's leading conjurors to his Paris laboratory to perform their magic under research conditions. In addition to closely observing their performances, Binet used a 'chronophotographic gun' invented in 1881 by Étienne-Jules Marey. This had a grip, a barrel containing the camera lens, and a drum containing not bullets but photographic plates. These could be rotated at high speeds, with the plates exposed by squeezing the 'trigger'. Binet used the 'gun' to capture the magician's swift hand movements when shuffling or dealing cards. By analysing these images frame by frame, he could separate 'brute sensation from mental interpretation' to discover how their misdirection worked.[9]

Jastrow, another 19th-century psychologist who took a keen interest in magic, is a fascinating character in the history of early psychology. Although little known today, even by members of his profession, Jastrow was once America's most famous psychologist. He was born in Warsaw, Poland, on 30 January 1863, but his parents, Marcus and Bertha, moved the family to Philadelphia when he was three. A bright and studious child, Joseph became fascinated by psychology from an early age and was the first American to receive a doctorate aged only 20. Appointed a lecturer in psychology at the University of Wisconsin, he went on to create America's first laboratory specialising in the senses. His studies of optical and psychological illusions made him world-famous.

Much to the disapproval of his academic colleagues, he wrote numerous magazines and newspaper articles on pop psychology, appeared on radio programmes and authored many forerunners of modern self-help books, with titles including *Piloting Your Life*, *Effective Thinking* and *Sanity First*. In addition to journalism, Jastrow gave hugely popular public lectures and demonstrations, during which he conducted psychological tests on those attending.[10]

A round-faced, balding man with brown eyes behind pince-nez spectacles, Joseph was opinionated, arrogant and argumentative. Vehemently opposed to belief in the supernatural, he entered into furious rows with anyone promoting spiritualism. Among those with whom he had long and rancorous exchanges of letters were Sir Arthur Conan Doyle, creator of the Sherlock Holmes novels, and *Huckleberry Finn* author Mark Twain. It may have been from a desire to counter human gullibility and demonstrate that there is no such thing as 'real magic' that he invited two of America's most acclaimed magicians, Harry Kellar and Alexander Hermann, to his laboratory.

It was Kellar, nicknamed the 'Dean of American Magicians', who inspired the wizard in the 1939 musical *The Wizard of Oz*. One of his most popular illusions, performed before sell-out crowds, was the 'Levitation of Princess Karnac', in which he 'floated' a scantily clad young woman high above the stage.

French-born Alexander Hermann, whose intense eyes, goatee beard and imposing moustache made him everybody's idea of a real wizard, was a naturalised American who enthralled worldwide audiences by combining magic and comedy.

Flattered by the noted psychologist's interest in their art, both accepted Jastrow's invitation and allowed him to test their mental powers and manual dexterity. Had he hoped to discover that they possessed unique abilities, the psychologist must have been disappointed. The cognitive ability and physical prowess of these famous magicians were, he discovered, no better and sometimes slightly worse than the population as a whole.[11]

The truth is that a magician's speed and agility have little to do with their 'magical' ability. Most tricks are performed at an average pace, with the magicians using misdirection to achieve 'invisibility'. Recent vision research has established that we are only consciously aware of a tiny part of the sensory information potentially available to us. Magicians have known this for centuries and have acquired vast knowledge about the most effective ways to manipulate their audiences' attention to create and control their blind spots.[12] In their book *The Secret Art of Magic*, Eric Evans and Nowlin Craver compare magic to warfare, which they claim relies on the same principles of deception and misdirection.[13]

To understand why the human brain is so susceptible to deception and misdirection, we need to examine the 3lb organ that accounts for about 2 per cent of body weight, consumes 20 per cent of energy and runs on glucose at 12 watts.

The Meaning Machine

Such is the tsunami of incoming information confronting us daily that processing it in its entirety, even as little as 60 seconds' worth, would take a human brain over 800 years. To make sense of the world, with a minimum expenditure of time and energy, our meaning-seeking brains[14] constantly and automatically discard anything considered, at that moment, to be unimportant. As a result, of the 400 billion data bits per second clamouring for conscious attention, the brain discards all but around 20. Made without conscious awareness, this process can have a lasting good or ill effect on our lives.[15]

Our brain's ability to prioritise and filter information based on relevance, emotional significance and other factors means there's limited awareness of the vast amount of data the brain receives. Attending to one part of the visual field comes at the cost of paying attention to other parts. This unavoidable neglect is the root cause of inattentional blind spots.

'Despite this enormity of information and despite the rich visual impression we usually enjoy of the world, it is a common experience that visible events are overlooked when attention is paid to an alternative task,' says psychologist Ula Cartwright-Finch. 'Conversely, when attention is absent, it appears that our visual representation of the rest of the world is surprisingly limited.'[16]

Attentional limitations have intrigued thinkers since before the birth of Christ. In the 3rd century BC, the Greek philosopher and polymath Aristotle asserted that we could only attend to one thing at a time. 'Persons do not perceive what is brought before their eyes,' he wrote, 'if they are at the same time in deep thought or in a fright or listening to some loud noise.'[17] This view held until the 1st century BC when Roman educator Marcus Fabius Quintilianus cited the case of harpists who sang and played their instruments simultaneously to argue that it is perfectly possible to pay attention to several things at once.[18]

The 19th-century Irish astronomer, physicist and mathematician William Rowan Hamilton attempted to determine the extent to which this was possible by tossing a handful of marbles on the floor and asking himself how many he could see simultaneously. Finding it 'difficult to view more than six', Hamilton concluded that blind spots were more likely to occur when a person attempts to simultaneously pay attention to more than six objects, situations or ideas.[19]

As the world becomes increasingly complex and the demands on our ability to pay attention become ever more significant, understanding what we can and cannot see becomes more urgent. Take, for example, our inability to see more than one thing at a time.

Look through a window at dusk and, provided light intensities are suitably balanced, you can choose, at will, whether to look at what is happening outdoors or at the room's interior reflected in the window. What you can never do is see both simultaneously. Observe one, and the other vanishes.

In the late 1960s this everyday experience inspired psychologist Paul Kolers to develop one of the earliest studies of looking without seeing. He devised a headgear fitted with a half-silvered mirror. Theoretically, it would allow him to see everything ahead of and behind him simultaneously. As with a window at dusk, although both images were potentially visible to him at all times, one always disappeared when he viewed the other.[20]

A few years later, Ulrich Neisser and Robert Becker concluded that while Kolers' studies were interesting, they needed to be closer to our everyday perceptual experiences to be of any significant value.[21] They showed their subjects videos of two different activities on separate monitors. The first was of two men playing the 'hand game'. In this, one player tries to slap the other's hands while the second man attempts to escape the blow by jerking his hands back or sideways.

The second featured three men tossing a basketball between them. Half-silvered mirrors were used to superimpose the hand-slapping game over the one of the ball being passed. The Cornell University student observers were instructed to press a switch at each successful hand slap and ball pass. Results revealed that subjects could not keep track of both activities, failing to notice hand slaps or ball passes.

Ten years later, psychologists Robert Becklen and Daniel Cervone conducted a similar, if more technically advanced, study with the same purpose.[22] They superimposed videos of two teams, one wearing white shirts and the other black, passing a ball to one another. The observers' task was to count the number of passes by the black-shirted team while ignoring passes by the white-shirted players. The task was complicated by their 'ghost-like' appearances and the fact that they walked through one another whenever they overlapped on the screen.

After 25 seconds of play, a woman carrying a large white umbrella, partially covering her face, entered from the left. Taking no part in the game, she strolled between the camera and players before disappearing from view. With their attention focused on counting the passes, seven out of ten onlookers failed to see the woman, even when looking straight at her.

This idea was later taken up by Christopher Chabris and Daniel Simons of Harvard, who replaced the umbrella-carrying woman with someone dressed as a gorilla, who strolled through a ball game, invisible to those attempting to count the number of passes. The figure is almost impossible to miss once you know he will appear, but nearly half of those watching the video clip for the first time never saw the bizarre intruder.[23]

These and similar experiments demonstrated that we can only focus on one thing at a time. This raises the question of how the brain decides what to attend to and what to ignore. We need to consider attention in greater detail to provide an answer.

Paying Attention

Making sense of our surroundings starts with a pre-attentive stage, during which our senses gather information about the world around us, supplementing it through our long-term and short-term memories. Long-term memory represents all we have seen, heard, smelled, tasted or touched. While it seems reliable and accurate, as we describe in later chapters there are many ways in which such memories can be changed, distorted or even implanted.

Short-term memory has several limitations that can lead to blind spots. The first is the brief amount of time, around 30 seconds, over which it can store new information. Unless refreshed or transferred to long-term memory, any memories held there will vanish forever. Second, it has limited storage capacity. For decades, this was said to be the reason why US phone numbers contained only seven digits; or up to six in some other countries.[24] Some formal studies have disputed this number, suggesting we can hold no more than four items in short-term memory.[25]

Psychologists Nelson Cowan and Candice Morey liken our ability to hold things in short-term memory to painting a canvas of limited size – which represents our working memory – with rapidly drying paint – our fading sensory memory. 'The number of objects that can be painted onto the canvas depends on both the size of the canvas and the time available before the paint becomes too dry to use,' they suggest.[26]

Every fresh memory overwrites older ones, making it hard, often impossible, to accurately recall what we saw moments earlier. For example, drivers' recall of recently seen road signs is remarkably poor, probably because their memory of seeing the sign has been overwritten by subsequent information.[27]

You can try this yourself using a test designed by Dr Don Read. While counting down from 100, also count the number of 'F's in this sentence:

Finished files are the result of years of scientific study combined with years of experience.

Read found that, when reading it for the first time, between 85 and 90 per cent of people say there are three 'F's when there are, in fact, six. 'Many discover they must read it numerous times to detect more than three "Fs",' he reports. 'In some cases, the remaining "Fs" are never discovered. I have witnessed colleagues and students becoming incensed when assured that six can be found.'[28]

Attention plays the part of an adjudicator in an unrelenting competition for neural resources. Before any conscious effort can be made to attend to one thing rather than another, your brain has already unconsciously decided in the unending flow of incoming sensory signals what to attend to and what to ignore.

Imagine reading an absorbing novel while seated at the window of a city coffee shop. Despite potential distractions from the endless flow of cars, trucks, passers-by and fellow customers, you can easily focus on the book. Suddenly, a small, black beetle crawling across the tabletop catches your eye. The book is forgotten; you watch in fascination as it disappears behind your mug.

Which poses an interesting question. 'How,' asks neuroscientist Radek Ptak, 'did your brain determine that the insect was more relevant than any other surrounding stimuli?'[29]

As mentioned earlier, although we perceive massive amounts of information, our brain's capacity to process all this information is far more limited. From endless streams of incoming signals, it has to select only those deemed most important at any given moment. In the case of visual information for instance, movement, like the beetle starting its crawl across your table, catches our attention due to its inherent survival implications. Things that moved were more likely to pose a risk to our evolutionary ancestors than those that remained stationary.

In his *Endeavour Journal*, Joseph Banks, chief scientist on Captain Cook's first voyage of discovery, described Aboriginals' indifference to their arrival off the coast of the southern country later known as Australia. 'Some people were seen. Not one was observed to stop and look towards the ship; they pursued their way in all appearance, entirely unmoved by the neighbourhood of so remarkable an object as a ship must necessarily be to people who have never seen one.'

A short while later, the *Endeavour* dropped anchor opposite a small village. 'Women and small children were on the beach,' wrote Banks. 'They often looked at the ship but expressed neither surprise nor concern. In a while, the four fishermen returned to the village, hauled up their canoes and went about their business, again totally ignoring the *Endeavour*, which was anchored just half a mile away.'[30]

One reason for their lack of reaction to the ship could have been an inattentional blind spot. In the second instance it might also have been that since the islanders didn't consider a motionless boat to pose any danger to them, they paid no attention. Only when the crew moved towards the land was any interest taken.

This raises the question of whether, when your gaze switched instantly from book to beetle, the shift of attention was voluntary or involuntary. After all, you could easily ignore the distractions of the street while paying attention to the novel.

Current research on the brain regions responsible for visual attention is dominated by a debate as to whether we can select a stimulus solely based on its characteristics (colour, shape and so on) or whether behavioural predispositions may overwrite stimulus-driven attention capture.[31] If, for example, we are performing an important task, it seems likely that our attention will be focused more on objects relevant to that task (the book in our previous example) than on the colours and shapes of other things (the traffic outside the window). At the same time, anything with a unique colour and shape, such as an exotically coloured butterfly fluttering through the window, will likely catch our attention no matter how focused we are on the task.

During the 19th century, German physicist and physician Hermann von Helmholtz suggested attention was like directing a spotlight to illuminate a small portion of our surroundings while leaving the remainder in obscurity.[32] Much research and theorising on attention in the past 50 years has concerned the observer's ability to decide how that spotlight should be directed – which parts of the world around us should be flooded with light and which are left in darkness. Using a single beam of light rather than attempting to illuminate the entire scene makes biological sense since it avoids squandering scarce metabolic resources on images of objects without bearing on one's current or future actions.

Since predicting which objects will become relevant and where they will be is often impossible, we use several visual functions to prioritise the most significant of the ones around us. Eye movements enable us to centre critical parts of our surroundings on those regions of the retinas with the greatest resolution. As a result, aspects of our surroundings we believe to be most important are pin-sharp, while everything else remains slightly fuzzy. Selectively increasing sensitivity to the features of potentially relevant objects, such as their orientations, motion directions and colours, enables them to stand out from everything else.[33]

If we assume that paying attention involves a single mental faculty rather than several, the process should stay the same irrespective of what we see. There should be an economical way to select from the torrents of sights, sounds, tastes, touches and aromas we experience every waking second of every day. This raises a further question. If attention is like a 'spotlight', which part of our brain determines where it shines?

The mechanisms controlling this 'spotlight' are either bottom-up or top-down. Bottom-up attentiveness originates with the stimulus, such as that black beetle crawling over your tabletop. We pay attention to a specific aspect of our surroundings because what we look at compels us to do so, for example, by being unique, unexpected or fast-moving. If creating a video for social media, for example, your first words or images should arouse curiosity and make people eager to know more.

Top-down attentiveness is a conscious and voluntary action in which you purposely seek out a specific part of your surroundings. While reading your book, top-down attention enables you to consciously focus on the text. When the beetle appeared, involuntary, bottom-up attention momentarily took over. If, fascinated by the insect, you continued to study it, then top-down attention would have regained control.

Now imagine that, rather than noticing the beetle while sipping coffee, you were an entomologist scouring a patch of ground for a rare species. In this case, you are likely to ignore all insects except the one you seek, employing top-down attention to distinguish between them, perhaps by size, colour or shape.

'In deciding what to focus on, the brain essentially shines a spotlight from place to place,' says Dr Jeremy Wolfe of Boston's Visual Attention Lab. 'A rapid, sweeping search that takes in maybe 30 or 40 objects per second. Our spotlight of attention is grabbing objects at such a fast rate that introspectively, it feels like you're recognising many things at once, but the reality is that you only accurately represent the state of one or a few objects at any given moment.'[34]

Whether a bottom-up or top-down mechanism is employed, the visual system can focus on only one or very few objects at a time, and anything lying outside that relatively small region of interest is ignored.

The Price of Paying Attention

As most people know from experience, concentrating hard on any activity is mentally and physically draining. As long ago as 1882, the Austrian physiologist Sigmund Exner wrote that he found paying close attention for an extended period 'in the highest degree exhausting'. After conducting some experiments in which he was concerned to ensure his results were 'as uniform as possible', Exner found himself covered with sweat and excessively fatigued 'although I had sat quietly in my chair all the while'.[35]

Just as your muscles will burn more energy when running a marathon than while taking a stroll in the park, the brain consumes energy reserves slightly faster when active; the difference is about 5 per cent more when struggling to solve a complex problem than when chilling out with a TV soap. Even quiet, reflective thinking will burn around 320 calories daily. A person who spends eight hours on mentally challenging work would only use about 100 more calories than if they were watching TV or daydreaming for the same amount of time.[36]

Psychologists Nilli Lavie and Jan de Fockert used brain imaging to establish the amounts of energy used by different regions and how these change as tasks become more demanding.[37] In one study, subjects searched for circles among diamond-shaped distractors. By pressing a key, they had to indicate whether a line segment in the circle was horizontal or vertical. Brain images obtained as they tackled this task showed increased activity in the frontal regions, the area above the eyes, suggesting additional demands on short- and long-term memory.

While thoughts, however profound, cost little energy, the bodily machinery necessary for supporting those thoughts is expensive. Most of your brain's energy consumption is used to sustain alertness and monitor the environment for important information. Managing other essential activities adds to the unending drain on resources. As energy-sustaining glucose dwindles, the brain's ability to stay on task tapers off and one can no longer sustain the same level of cognitive performance.

And this reduction in attention can have serious real-world consequences, with accidents caused by inattentional blindness being far from unusual. Public safety researcher John Treat and his colleagues report inattentional blindness has been responsible for numerous disasters at sea and in the air[38] and is be the sole or contributing factor in over 90 per cent of road accidents.[39]

On a November night in 1993, the 35-year-old teacher Eleanor Fry was driving a minibus loaded with a dozen students back from a concert in London's Albert Hall. Distracted by removing her glasses, she collided with the back of a motorway maintenance lorry. The minibus burst into flames and exploded. Eleanor and all but two young passengers perished in the inferno.[40]

United Flight 173, from New York's JFK to Oregon's Portland International Airport, had been smooth and uneventful. Then, as the crew began their final approach on a bleak December afternoon in 1978, an indicator bulb signalling the landing gear had lowered failed to light. Concerned that their wheels weren't down and in the right position, the three-person team aborted the landing and requested a holding pattern. For the next hour, they circled Portland, trying to determine what could have gone wrong. Captain Buddy McBroom and first officer Rod Beebe were so distracted they failed to watch their fuel gauges. The plane ran out of fuel and crashed in a built-up area 10km south-east of the airport. Eight passengers, two crew members, a flight attendant and the engineer died. The landing gear was found to be fully extended and locked in place.[41]

In a similarly tragic marine disaster, on 9 February 2001 a nuclear submarine, the *USS Greeneville,* sailed from Pearl Harbor to take a group of VIPs on a cruise under the Pacific. A few hours later, Commander Scott Waddle ordered 'an emergency main ballast tank blow'. Why he did this remains unclear, although some have suggested he may have wanted to impress his VIP guests by adding drama to an otherwise uneventful trip.[42]

The 6,000-tonne sub surfaced violently beneath the keel of a Japanese trawler, the MV *Ehime Maru.* The training vessel, carrying teenage students and their teachers, was sliced in half and sank almost immediately. Three crew and nine students drowned. A subsequent naval investigation discovered that the *Ehime Maru* was visible on the *Greeneville*'s sonar, and that Scott Waddle had been looking straight at the boat through his periscope, yet he had failed to see it or hear the urgent sonar pinging warning of a vessel directly above them.

Inattentional blind spots have also led to loss of life in a wide range of natural tragedies. Dr Michael Kinsey, a senior fire engineer based in Shanghai, states that whether people live or die in a burning building often depends on whether they pay attention to exit signs or are distracted by less relevant features in their surroundings. In one case, a man working on his computer was so focused on this task that he failed to notice the smoke circling his office until it was almost too late.[43]

Severe and widespread as they are, inattentional blind spots are not the only or even the most serious example of how looking does not necessarily equate to seeing. As explained in the next chapter, we often fail to notice changes in our surroundings, even when these are significant and, once detected, all too obvious.

You can put this to the test by choosing one of the six playing cards below and remembering it. We will identify that card later with a trick illustrating the second most common reason for failing to see what we should have seen: a blind spot first identified during Hollywood's era more than a hundred years ago.

Figure 2

CHAPTER TWO
Now You See It – Now You Don't

On a crowded street, you are stopped by a stranger asking for directions. While explaining the route he should take, two men carrying a door walk between you. Once they've gone past, you finish your explanation, at which point the stranger makes a surprising admission. He's not the same person who hailed you only moments earlier. As the door obstructed your view, he swapped places with one of the men carrying it.

Studying his appearance more closely, you realise he is right. The man who stopped you was taller, with a broader chest and darker hair. Once you have noticed these differences you cannot understand how you could have been so blind to such obvious changes.

First conducted more than twenty years ago by psychologists Daniel Simons and Daniel Levin, this study has been widely replicated, always with the same results. Fewer than half of those stopped noticed they were talking to two different people. Most continued giving directions as if nothing had happened. 'They were quite surprised to learn that the person standing before them was different from the one who initiated the conversation,' reports Daniel Simon.[1]

What is striking about change blindness is how glaringly obvious the change becomes once pointed out. Take, for example, the two photographs on the next page. While at first glance they may appear identical, there is a significant difference between them. Can you discover what it is?

If you spot the change, congratulations. Nine out of ten people taking this test fail to do so. Yet once you know what it is, you will never miss it again. You can find the answer at the end of this chapter.

Figure 3

In another test, we made four changes to a Las Vegas street scene. To make detecting these more difficult, they were not presented side by side, as shown below, but as a movie in which the images rapidly alternated in what is known as a 'flicker test'.[2]

Figure 4

When viewing the pictures side by side, attention must constantly move between them to spot the changes. When the pictures alternate, the brain has to register and record each change as it occurs, testing short-term visual memory to its limits.

To try this test, go to lewisandleyser.com/flicker-test or scan the QR code.

Did you spot all the changes? In our study on social media, only a tiny percentage noticed all of them.

And if changes to still images are hard to detect, how much harder is it to spot them with moving images? To find out, we created a 30-second video depicting a modern living room with a large cuddly toy and asked our viewers whether they thought it cute or creepy.

Figure 5

The question, however, to which 60 per cent said cute and 40 per cent creepy, was merely a distraction. What interested us was whether, with their attention focused on the toy, anyone would notice the ten changes we had made. None of the 15,000 who watched this video did so. To see the video, go to lewisandleyser.com/cute-or-creepy, or scan this QR code:

Another aspect of change blindness is known as 'satisfaction of search', where people stop looking for further differences once they believe they've found them all. Satisfaction of search (SOS) can have serious real-life consequences.[3] Radiologists, for example, are less likely to notice further abnormalities after detecting one on an X-ray.[4] Having spotted one illegal item, such as a bottle of water, in a passenger's luggage, airport security staff can easily overlook another. We will discuss SOS's dangers to airline safety later in this chapter.

How Hollywood Discovered Change Blindness

Failure to notice change was first observed not in the austere surroundings of a university psychology department but amid the glitter and glamour of Hollywood.

'Filmmakers wrote about change blindness ... well before systematic empirical work on visual integration in the cognitive psychology literature,' comment Simons and Levin. 'Moreover, their ideas about the sketchiness of visual memory often predated similar claims in the psychology literature.'[5]

In the early years of movies, boxing proved a popular subject, mainly for technical rather than dramatic reasons. Each two-minute-long bout could be recorded on a single 30m (100ft) film reel, the maximum length early cameras were capable of. Boxing also offered a second advantage in that, with all the action in a relatively small ring, the bulky, tripod-bound camera could remain stationary throughout.

This 'point and shoot' approach to filmmaking continued until directors such as David Wark Griffith found they could tell far more exciting and compelling stories by shooting from several camera positions and editing together different scene views. Griffith combined wide-angle shots, showing an entire scene, with close-ups of the performers' emotions. Previously, as in the theatre, film actors had been obliged to use broad facial expressions and physical gestures to express feelings and intentions; editing enabled them to be conveyed far more subtly.

The trouble lay in maintaining continuity. With the same scene being shot from different angles or on different days, it became all too easy for changes in an actor's appearance, costume or props to occur, sometimes in far from minor ways.

When directing *That Obscure Object of Desire*, Luis Buñuel originally cast Maria Schneider as his female lead. She proved so unreliable he replaced her with two other actresses, Angela Molina and Carole Bouquet. When editing the film, Buñuel alternated these two actresses across scenes. By the end, he was swapping between them in the exact location with only a few seconds' delay. Despite this, many in his audience were unaware of the change.[6] As film director Lev Kuleshov pointed out: 'When we … shoot the constituent parts of a scene at different times or insert a filmed element of one scene into another, we sometimes have to disregard small inconsistencies in the costume of an actor.'[7]

What surprised early movie makers was how few of their audiences noticed major continuity errors. This was especially true if a change occurred when an important character exited as another entered. The problem was that if the change went unseen by most of the audience, the same blindness could beset everyone making the movie. One of the most glaring examples of these occurs in Alfred Hitchcock's 1959 classic *North by Northwest*. In one scene, hero Roger Thornhill (played by Cary Grant) is confronted by Eve Kendall (Eva Marie Saint), a woman with whom he's falling in love. She draws a gun and shoots him at point-blank range. A small boy who hated loud noises was among the extras, seated near the two stars, and he can be seen sticking his fingers in his ears just *before* the gun is fired! And yet, despite the number of people working on the film, this went unnoticed and the footage made it through to the final movie.

Almost 40 years later, change blindness led to equally obvious continuity errors. In the 1995 neo-noir classic *The Usual Suspects*, a four-engine Boeing 747 is filmed coming in to land. When the same landing is seen from behind, the plane is suddenly transformed into a two-engine Boeing 767.

The problem of continuity becoming a victim of change blind spots was deemed so severe that, during the 1970s, leading Hollywood directors established a series of filmmaking conventions known as continuity editing rules. These rules were designed to prevent confusing cinema audiences by editing films in a way that is disorienting or uncomfortable to watch.

In a scientific investigation of change blindness in films, Simons and Levin produced videos in which props and costumes were changed between shots.[8] One shows two women sitting opposite each other at a crockery-strewn table. At the start, one wears a colourful scarf that disappears as the camera position changes, and in addition red plates become white, and the actors' hand positions alter between shots.

Cornell University students who watched these videos were asked whether any changes had occurred between one shot and the next. Only one claimed to have noticed, and he could not describe them. The students were asked to watch again and, this time, note down any changes they spotted. Even then, only two out of a total of nine were observed. The most frequently noticed change was the vanishing scarf, reported by 70 per cent of participants.

Simons and Levin felt that, while their findings confirmed the importance of change blindness, they needed to assess whether they had allowed too little time for them to be detected. They shot a second video featuring a man seated at his desk. When a phone rang in another room, he answered it. While no changes were made to costumes or props, the video did include one difference the researchers thought should be even more noticeable. The actor who got up from behind his desk differed from the one who answered the phone. Although he was constantly in view, only a third of those watching spotted the switch. All who watched the video could describe what the actor behind the desk looked like and how he was dressed without noticing that he became a different person between one scene and the next.

In another study, Simons tested whether, having detected one change, his subjects would find it easier to spot others.[9] Because his 'gorilla video', described in the previous chapter, had become so well known, he decided to use a subtle variation of the original. By filming it in front of a green screen, he could alter the background colour digitally. This started bright red when players began passing the ball, gradually evolving into gold as a man dressed in a gorilla costume appeared. At the same time, one of the black-shirted players walked off, remaining unseen for the remainder of the video. Afterwards, viewers were asked how many passes they had counted and whether anything unexpected had happened during the video.

Fewer than one in five of those who had previously seen the gorilla video noticed the background changing colour or that a black-shirted player had vanished. 'Although subjects who knew to look for a gorilla were much more likely to spot it, they were no more likely to notice the other unexpected events,' reported Simons. 'In fact, they might even be less likely.'[10]

Change Blindness and Gabor Patches

Psychologists sometimes use Gabor patches in their experiments to understand how and why change blindness occurs. Invented by Nobel Laureate Dennis Gabor in the 1970s, these patches consist of black and white bars that can be oriented in any direction.

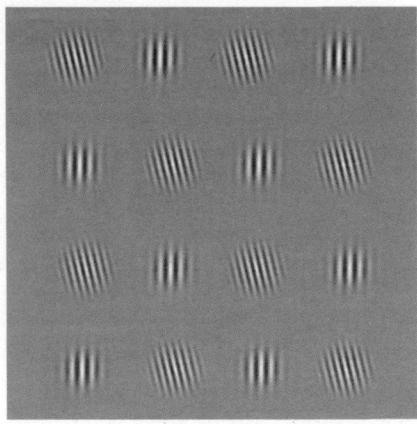

Figure 6

Psychologist Richard Yao and his colleagues asked subjects to watch a series of Gabor patches travelling down and across a computer screen. Their task was to spot a 15-degree rotational change in a randomly chosen patch moving in an L-shaped pattern. When moving down or across the screen, the change in a patch's orientation was detected 70 per cent of the time. When it occurred as the patch stopped travelling down and began moving horizontally detection rates fell to 14 per cent.[11]

Change blindness occurs for two main reasons: short-term visual memory limitations and neural fatigue. As with a muscle, the more frequently a neuron fires, the more tired it becomes and the less able it is to encode information correctly, for instance the orientation of Gabor patches. Short-term visual memory storage capacity can be likened to a short conveyor belt. Once all the available space is taken, an earlier item must drop off before a new memory is added. Where Gabor patches are concerned, research by Ronald Rensink indicates that the conveyor belt has room for up to five oriented-bar items. Once that limit is reached, earlier ones will be lost.[12]

Magicians have known about and exploited this for centuries. They use abrupt changes in the direction of movement to catch our attention, allowing them to conceal their method for performing a trick. Pickpockets, for example, will often move their hands in a fast, linear fashion, invoking attentional shifts in their intended victims. These victims rapidly shift their gaze from one location to another, unsure of where to look next.

Another example is the wave change card trick, in which one card is mysteriously transformed into another as the magician rapidly moves the pack around. In the illustration below, the ace of hearts has somehow morphed into the two of clubs. To see how effective movement is in creating change blindness and to learn the secret of how to perform the wave trick yourself, go to lewisandleyser.com/wave-change or scan the QR code.

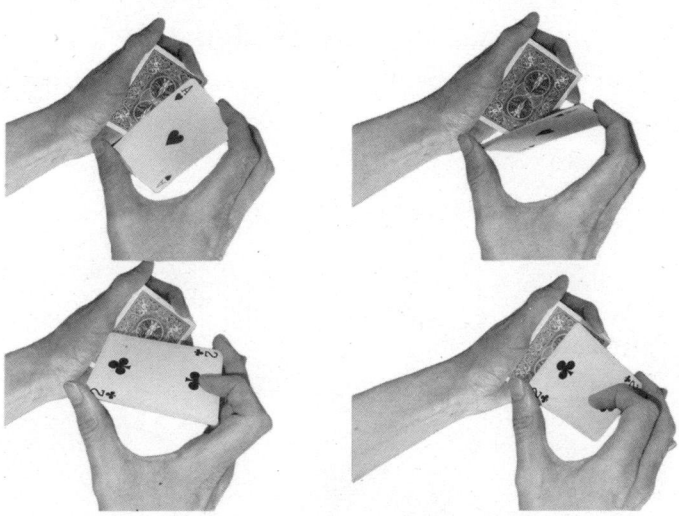

Figure 7

One theory proposed to explain change blind spots concerns what psychologists call our functional field of view (FFOV). This is the total visual field area from which useful information can be acquired without moving your eyes or turning your head.

Our ability to focus sharply on any specific feature in our surroundings is limited. Only one small area in each eye, a 0.35mm-wide depression known as the fovea, is capable of high visual acuity. You can test the limits of your sharp focus by staring at the black rabbit in the centre of the sequence below. Depending on your age, those on either side of the black rabbit will be increasingly blurred.[13] To see any of those further to the left and right, you must shift your eyes or turn your head to move the fovea over the image.[14]

The narrower someone's functional field of view, the less likely they are to spot changes at the periphery of their vision.

Figure 8

Researchers led by Dr Heather Pringle set out to discover whether certain aspects of a visually complex scene made changes close to the FFOV boundary more or less likely to be detected and identified.[15] Pringle showed volunteers, aged 18 to 80, pairs of photographs of street scenes taken from the perspective of a motorist looking through the windscreen. One picture in each pair had been modified to change the location of a building, the colour of a car and the absence or presence of a road sign. Subjects could search freely for the changes and were asked to respond as quickly and accurately as possible by pressing a button and describing what they had seen.

Pringle's studies demonstrated that whether or not a change is noticed depends on six factors:

Eccentricity: The distance a change occurs from the centre of vision, the region on our retinas where the eyes' ability to focus precisely is greatest.

Meaningfulness: The relevance that part of the scene has for each observer. For example, in Pringle's street scene study, a shop sign colour change would likely have had less meaning than in a traffic signal.

Saliency: How conspicuous a feature or location appears. The more easily a change is seen, the more likely it is to be seen.[16]

Categorisation: The brain uses several mental conservational shortcuts because our attentional reserves are limited. One of these is categorising sensory information by placing it into labelled boxes, the most important being 'like' and 'unlike'.[17] As an illustration of this, in the study described at the start of this chapter, the results differed significantly depending on whether the people seeking directions were *like* the direction giver (the switch was noticed 100 per cent of the time when all involved were students) or *unlike* them in terms of age and occupation (it was noticed only 35 per cent of the time when those seeking directions were dressed as construction workers).

 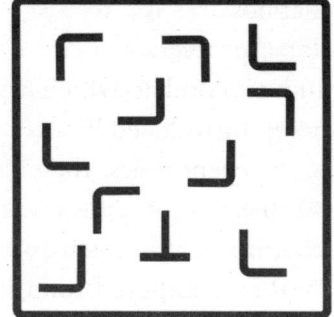

Figure 9

Prevalence: The more distractors there are, the harder it is to spot a change in the target item. You can understand the difference a high or low prevalence makes by finding a letter 'T' shape in the illustration below. When there are only three to choose from, the task is easy. The greater the number of distracting shapes, the trickier it becomes and the longer it takes.[18]

Age: Subjects aged between 18 and 33 proved somewhat faster at noticing changes than those aged 50 to 80. Younger subjects spotted them with 100 per cent speed and accuracy, while older ones only scored 92. Even when low-salience, relatively meaningless changes occurred on the boundary of the subjects' FFOV, younger subjects scored 90 per cent, while older ones scored 72 per cent.

Let's examine how one or more of these factors affects real-life situations.

Change Blindspots in Real Life

Change blindness is widespread, which can have serious, sometimes fatal, consequences. As seen earlier in the chapter, it can have a major impact in medical settings, where change blindness when reading X-ray images can lead to an incorrect diagnosis. As consultant radiologist Mark Randon told us: 'It's a problem which has dogged the profession for decades and to combat which rigorous new training programs are being introduced.'[19]

In recent years, the challenge radiologists face has become all the more complex with the introduction of new medical imaging technologies, such as magnetic resonance imaging (MRI), computed tomography (CT), and positron emission tomography (PET). Today, a single chest X-ray has been replaced by hundreds of images, every one of which must be scanned.

To test radiologists' ability to detect unusual changes on a CT scan, psychologist Dr Trafton Drew and his colleagues inserted the image of a small gorilla on images of lungs screened for cancerous nodules. 'If someone pointed at the correct location in the image and asked, "What is that?",' Drew explained, 'you would have no trouble answering, "That is a gorilla."'[20]

Yet despite looking straight at it, 20 of the 24 radiologists taking part remained blind to the gorilla. This was mainly due to the image's low saliency and lack of meaning to them. 'They don't see that it's a gorilla because they were looking for cancer nodules, not gorillas,' says Drew. When he conducted the same experiment with the lay public, no one spotted it. We obtained similar results when replicating this study by creating an X-ray containing a similar-sized gorilla image. To discover whether you can spot it, go to lewisandleyser.com/radiologist or scan this QR code.

It should be far easier to see the gorilla since, unlike the radiologists in Trafton Drew's study, you know there is an anomaly to look out for, but you'll find a guide to its location at the end of this chapter.

Another area in which change blindness can have serious consequences is in airport security. In an ingenious series of experiments, Professor Jeremy Wolfe demonstrated the challenges facing airport security officers scanning passenger luggage for prohibited items.

After concealing guns and knives in 21 carry-on cases, he had them scanned by the same X-ray machine used at airports and then showed the images to volunteers (none of whom were professional security officers, so the number of targets they miss remains unknown).

When weapons were concealed in one of only 20 cases, most were detected, but when hidden among 2,000 cases – a sample size closer to that which would occur at a busy airport – change blindness led to screeners missing around 30 per cent of them.[21] As Wolfe points out: 'If you often don't find it, you don't find it often!'

You can experience the challenge confronting airport security officials daily by visiting lewisandleyser.com/scanner or scanning the QR code.

There, you will see four pieces of luggage passing through a scanner. Each pauses for a few seconds, during which time you must determine whether any contain a prohibited item.

And it's not just security personnel in the aviation industry who face challenges due to blind spots. Consider pilots.

When we attempt to pay attention to everything, an approach often found among trainees and newly qualified pilots, critical changes may be missed. This is especially likely in low visibility or when flying through heavily congested air zones.[22]

Head-up displays (HUDs) – initially developed for the military but now increasingly found in commercial aircraft – enable pilots to view the primary flight instruments, such as altitude, airspeed, flight path and direction, while scanning the horizon for obstacles, weather, terrain, runways and other aircraft. When this information rapidly changes, as often happens, the pilot must quickly scan and remember the previous information before receiving the new information.

In a study of commercial pilots using the latest HUD, an alarming number of vital changes went undetected. During a simulation, experienced pilots had to decide whether it was safe to land under low visibility. Half the pilots remained blind to the presence of other aircraft or vehicles on the runway and, until stopped by the experimenter, continued to attempt to land the plane.[23]

Research has identified many reasons why there are more errors made when using HUDs compared to using flight instruments directly, including screen brightness (especially in contrast to a dark sky) and the pilot's seat position, which affect perception.[24] It's also been argued that too much information is displayed, reducing the pilot's limited attentional resources. These can either be narrowly focused, with all their attention concentrated on a single feature in the surroundings, or widely divided. Numerous features, some irrelevant to the current situation, are reviewed when using HUDs, with the pilot's gaze constantly flitting from one to another. Both too narrowly focused and overly divided attention increases the likelihood of change blindness, degrading pilot performance and raising the risk of error. Human factors such as this are responsible for more than 70 per cent of aviation accidents.[25]

A tragic case of too narrowly focused attention, very much like the cause of the United Flight 173 crash in Portland, Oregon, occurred in December 1972 when an Eastern Airlines Lockheed TriStar was about to land at Miami International Airport. As the flight crew lowered the landing gear, a light that should have indicated the nose wheel was down failed to illuminate. Informing the Miami control tower that he was abandoning the approach, the captain requested a holding pattern. As they circled, the flight crew continued focusing on the malfunctioning lamp while failing to notice their aircraft descending. By the time they paid attention to altitude, it was too late. In the crash that followed, over a hundred of the 173 passengers and crew died.[26]

A further context in which change blindness can lead to significant real-world consequences is in the military. Several studies by Paula Durlach of the US Army Research Institute have led the military to invest significant time and money in researching change blind spots when detecting aggressors.[27]

One study focused on the effect of navy combat information centres' computer monitoring. Operators are tasked with monitoring computer-displayed maps depicting a bird's-eye view of aircraft, sea vessels, ground installations (airports), and commercial air corridors. Each is shown as an icon of a different colour and shape. The operator's task is to track the movements of these icons and evaluate them as friendly or hostile. While doing so, he or she must watch another computer screen displaying alerts, notifications and questions.

In a study of the effects of change blindness on performance, Durlach and her colleague Laticia Bowens asked observers to report any icon changes in colour, shape, position, appearance or disappearance. Observers were also asked to perform different tasks, such as sending a text message to another unit.

When carrying out these tasks, participants had to negotiate menus and windows providing the required information, such as the text of their message and the addressee. These windows were superimposed upon (and blocked the view of) the map.

When the researchers caused an icon to change simultaneously with closing the window associated with the task, only half the changes were detected, compared with a detection rate of about 90 per cent under normal circumstances. What made this high failure rate especially concerning was that the operators had been told exactly what would happen. 'The participants knew they would be tested for change detection and the types of changes they would be tested on,' reports Durlach. 'Moreover, the screens were relatively uncluttered (only 8 contacts present at a time), compared with when the system is actually in use (50–100).'[28]

The study identified several factors contributing to whether or not a change in an icon would be detected, for instance yellow icons were more likely to be missed than blue, green or red ones.

Durlach and Bowens suggested several ways to prevent the user from relying on their memory of previous displays. For example, a 'snail trail' on the map could be laid down if an icon changes position, or multimodal interfaces could be introduced in which information is distributed across multiple sensory channels, primarily visual, auditory and touch.[29]

Your Chosen Playing Card

Remember the trick at the end of the previous chapter, with six playing cards? We asked you to pick just one of them and remember it. Look at the cards again; we have made your chosen card vanish.

Figure 10

After reading this chapter, you should be able to see how and why this happened. If not, you'll find the answer below.

Answers
Figure 3: The airliner in picture B is without its left (port) engine.
Figure 10: All five playing cards are different.

CHAPTER THREE
It's Your Choice – Or Is It?

A young man is shown pictures of two attractive women and asked to choose the one he would most like to date. He is then handed the photo again and invited to explain what appealed to him about her. He does this, enthusiastically praising her features and expression. What he fails to notice is that the images have been swapped. If he selects woman 'A', he is asked to extol the virtues of woman 'B', and vice versa. To watch this in action, go to lewisandleyser.com/choice-blindness or scan the QR code.

While it might seem unlikely that anyone would mistake one person for another, this is precisely what psychologists Petter Johansson and Lars Hall found to be the case.[1] Seven out of ten people failed to notice the deception and confidently expressed reasons for their choice. 'There was much more spark in her eyes,' one man told them. Another commented favourably on her 'big eyes', adding, 'She's got a nice mouth, very shapely, I think.'

A similar result was found among women to whom pictures of men had been shown. 'He looks more pleasant,' one woman explained, while another commented on how 'kind' her unchosen man looked and how much he reminded her 'of a good friend'.

In another study of facial preferences, pairs of male and female faces were shown, and participants were asked to rate them on a scale from 'very unattractive' to 'very attractive'. Presented a second time with one of the photos they'd rated as 'very attractive', participants were asked to select the features that determined their choice by pressing one or more of seven buttons. Six were labelled with facial characteristics such as 'mouth', 'proportion', 'eyes', 'skin', 'nose' and 'shape' while the seventh offered them a chance to change their minds and choose the other face.

Some of the faces offered on the second choice were swapped and one group of participants was warned that the face they saw on the second occasion was not necessarily the one they had chosen on the first. A second group was not alerted to any potential switching. Even among the first group, who had been alerted to a possible manipulation, 24 per cent failed to notice. Studies such as these demonstrate a visual failure known as choice blindness in which people fail to notice that their choice has been manipulated or switched. They not only fail to detect the switch but typically confabulate reasons to justify the choice they didn't actually make and remain unaware of the inconsistency.

Intrigued by Johansson and Hall's findings, Anna Sagana and her colleagues examined how choice blindness affects the reliability of eyewitnesses' facial recognition. Posing as tourists, researchers stopped passers-by on a busy city street and engaged them in a short conversation. Not long after, the same unsuspecting passers-by were stopped a second time and asked to identify the experimenters from a line-up of photographs. On a third occasion, they were stopped and shown a photograph of the person they'd identified and asked to explain their identification. This time, however, the picture they had initially chosen had been swapped for another and 68 per cent failed to notice the change.[2]

While these experiments don't have any impact on real-world situations, it's important to note that choice blindness can, in some situations, have severe consequences for those incorrectly identified. Nowhere more so than in a court of law.

Before DNA testing and advanced forensic techniques, the police extensively used identification parades to confirm a witness's identification of a suspect. Suspects were placed among a row of similar-looking men or women recruited off the street and witnesses walked down the line, studying each face to see whether they could make an identification.[3]

Known as 'dock identification', this practice was responsible for numerous wrongful convictions, with a US study finding errors in eyewitness identification to have been the critical factor in 1,000 convictions of innocent persons. Similarly, America's Innocence Project reported that mistaken identification had led to wrongful conviction in 87 per cent of cases where the accused were later exonerated by DNA evidence that had been unavailable at trial.[4]

Over a century earlier, such a fate had befallen British businessman Adolf Beck, who, nine days before Christmas in 1895, walked out of his London hotel – and found himself heading straight to jail. On leaving his lodgings, he was accosted by a middle-aged schoolteacher named Ottilie Meissonier. She accused him of being a confidence trickster who had robbed her of valuable watches and jewellery. When the police arrived to see what the commotion was all about, she told them that, a few days earlier, Beck had engaged her in conversation under the pretext of mistaking her for another woman. The middle-aged man, who had called himself Lord Willoughby, had seemed so pleasant and respectable that she had agreed to have afternoon tea with him. After several meetings, and being short of cash, Ottilie had sold him her watches and jewellery but the cheque he had given her had bounced. Both 'Lord Willoughby' and her jewellery had disappeared.

Beck denied having ever met Ottilie Meissonier before that day and readily agreed to an identification parade. Before this took place, however, an inspector assigned to the case discovered that, over the previous two years, 22 other women had also been defrauded by a man calling himself Lord Wilton de Willoughby. All attended the parade, and each positively identified 54-year-old Beck as the man who had defrauded them.

After the identification parade, things looked bad for the unfortunate Adolf Beck. They were soon to become even worse. After further enquiries, the police concluded Beck was actually a man named John Smith, who had previously served five years for defrauding single women using the name Lord Willoughby. At Beck's trial in late 1895, the officer who had arrested Smith, twenty years earlier, told the court. 'The prisoner is the man. There is no doubt whatever. I know quite well what is at stake on my answer and say that he is the man.'[5]

Based on eyewitness testimony alone, Beck was pronounced guilty of ten misdemeanours and four felonies. The judge, who had also presided over John Smith's trial, sentenced him to seven years of penal servitude. In jail, he was assigned John Smith's original prison number and branded with the letter 'W' denoting a repeat offender.

For the next two years, Beck's solicitor petitioned the Home Office, without success, to have the case re-examined. Then, in 1898, an official reading the files noticed that John Smith, a Jew, was circumcised while Beck was not. In the face of this irrefutable fact, the State accepted that Adolf Beck could not be John Smith. But they also decided his sentence for robbing Ottilie Meissonier should stand. The 'W' was removed from his prison record, but he remained imprisoned until 1901 when he was paroled for good conduct after six years behind bars.

Beck must have thought his long nightmare had come to an end. Then, three years later, another six women levelled the same accusation against him. As before, he was picked out from a police line-up. Once again, a jury found him guilty based solely on eyewitness evidence. Fortunately for him, the judge felt unhappy about the case and postponed sentencing despite intense pressure from the Home Office. Within the week, the judge's unease over the trial was proved right. John Smith, whose real name was Wilhelm Meyer, was arrested, brought to trial and jailed for five years. After a public outcry, Beck received a pardon from the king and £5,000 (£600,000 in today's money) in compensation for his false imprisonment.

Cases such as this illustrate the difficulties people have not only in identifying a face seen only briefly and, often, under severely stressful circumstances but also how easily choice blindness leads to one person's face being mistaken for another.

While memory-based eyewitness identification is fallible, however, is one based on photographic evidence more reliable? In recent years, CCTV has increasingly been used for identification purposes but, far from being foolproof, research has shown it to be a highly unreliable way of identifying suspects.

In one study, participants watched a low-quality, black-and-white CCTV video of two men taking part in an armed robbery.[6] After seeing the footage, they were handed several photographs of similar-looking men and asked to identify the culprits. Only a minority of participants were correct in their choices. Even when the same scene was viewed on broadcast-quality video, an incorrect identification was made on over 36 per cent of occasions.

However, choice blindness doesn't only occur in situations with a criminal aspect, and, intrigued by the findings from their 'swapped faces' study, Lars Hall and his colleagues decided to see whether choice blind spots were unique to human faces or extended across other aspects of life. While choice blind spots might occur when dating, they theorised they would never do so when eating. As Lars Hall points out, 'Consumers often have firm opinions about marketing and branding of products as such, and they think and reflect about how these factors influence their own decisions.'[7]

To put this belief to the test, they invited shoppers at a local market to sample two different varieties of jam and tea

Figure 11

and choose which alternative in each pair they preferred. Immediately after picking, shoppers were asked to test the jam or tea again and explain why they had made that choice. Unknown to them, the contents had been switched using a jar with two lids and an internal partition.

As a result, the jam or tea they tasted the second time was very different from their first choice. Spicy cinnamon apple might be swapped for bitter grapefruit, or mango for aniseed, and yet seven out of ten consumers failed to notice these dramatic changes in taste or smell between the first and second tasting.

'In the great majority of trials, they were blind to the mismatch between the intended and the actual outcome of their choice,' reports Petter Johansson. 'They instead believed that the taste or smell they experienced in their final sample corresponded to their initial choice.'[8]

If choice blind spots can be present in such basic human desires as finding a partner and eating, do they also occur when we turn our minds to higher things, such as appreciating beauty? To find out, Johansson and Hal showed subjects pairs of abstract patterns and asked them to choose the most aesthetically appealing. The image was then presented for the second time. And, as in their previous experiments, participants were asked to say what about the pattern appealed to them. As before, images were randomly swapped, so some participants were now looking at the picture not chosen initially.

'Surprisingly, the subjects seldom noticed the switch,' reported the researchers, 'and often remembered the manipulated choice as their own. Combined with our previous findings, this result indicates that we often fail to notice changes in the world even if they have later consequences for our actions.'

In the twenty years since Petter Johansson and Lars Hall first coined the phrase 'choice blindness', researchers have found it occurring in fields as diverse as finance planning, legal decisions when purchasing homes, computers, and mobile phones, and, most worryingly of all, deciding which way to vote in a general election.

In a study entitled 'Lifting the Veil of Morality', Hall and Johansson asked participants their beliefs regarding moral dilemmas receiving attention in the media. They might be asked, for example, to say whether or not they considered it 'morally *defensible* to purchase sexual services in democratic societies where prostitution is legal and regulated by the government' (italics added for emphasis).

When shown these for a second time, some statements were reversed. The statement above, for example, now asked whether it is: 'morally *reprehensible* to purchase sexual services in democratic societies where prostitution is legal and regulated by the government' (italics added for emphasis).

Blind to their original choice, with the same enthusiasm as before, around half the participants endorsed a statement they had not made, demonstrating the extent to which moral attitudes are malleable and relatively unstable.[9]

Similarly, in a study of choice blindness in financial decisions, Owen McLaughlin and Jason Somerville showed subjects reports outlining the investment options offered by one of six fictional pension companies and asked them to make a choice. Each option was described in relatively simple language, and the associated risk ranged from low to high.

In some trials, the participant's chosen portfolio was then presented to them for comment with no changes, but in others, experimenters replaced the selected portfolio with one containing a significantly different risk profile. They found that between 60 and 70 per cent of manipulated trials went undetected by participants, and even with the most detection-favourable conditions, 50 per cent went unnoticed.[10]

Taking this further, Jessica Schanzer found choice blindness extends not only into our personal finance choices but to almost every purchase we make. She ran a study asking people to choose between two different options for a consumer good. Each participant was either asked to choose between two different laptops, two different cars, or two different televisions. They were given the price and a description of other key attributes for each. Those choosing between two different laptops, for instance, were given information on the display, resolution, memory, hard drive and battery life. Schanzer then switched either the item's price or other attributes before asking participants why they had chosen that item over the other: 71 per cent of those taking part remained blind to her manipulations, confidently and enthusiastically giving reasons for a choice they never made.[11]

Indeed, one of the first studies looking at the hidden factors that force our choices was on consumer goods. Richard Nisbett and Tim Wilson asked shoppers to choose the best quality tights from a selection of four and explain why they made their choice. Even though the only difference between them was their position on the counter, participants offered justifications for their choice, such as a superior knit, greater sheerness or more elasticity. None realised their choice had been forced by what is termed 'the position effect' whereby items placed to the right are typically preferred over items to their left.[12]

Research by cognitive psychologist Hiroshi Nittono at Osaka University suggests we gravitate towards the right side when making selections, because between 70 and 90 per cent of us are right handed. This right-side preference is especially true when viewing a display or choosing between options.[13]

Conformity

Until now, we have only considered situations in which people were unaware of their choice blindness until it was pointed out to them. There are occasions, however, when we consciously choose not to see.

From school days onward, people's preferences are greatly influenced by a desire to look, think, believe and behave in ways that conform to the expectations of those around them. Defined as 'yielding to group pressures', conformity can take different forms. It may be brought about either by a desire to 'fit in' or be liked (normative), to be correct (informational), or to conform to a social role (identification).

One of the first psychologists to study conformity was Arthur Jenness. In 1932, he asked participants, individually, to estimate how many beans there were in a bottle. He next invited a group to provide an estimate through discussion. Jenness then interviewed the individual participants again and asked if they would like to change their original estimates or stay with the group's estimate. Almost all changed their guesses to be closer to the group estimate.[14]

A similar early demonstration of normative conformity came with a 1951 study by American psychologist Solomon Asch. He assembled a group of six students, of whom five were actually his accomplices and only one an actual participant, naïve to the purpose of the experiment. All were shown drawings of different-length lines and asked to decide which was the longest. Asch's hypothesis was that a minority will conform to the majority view even when it contradicts the evidence of their own eyes. He found that when the five accomplices insisted a short line was the longest, the sixth agreed, rather than contradict what they believed to be the majority opinion.[15] 30 years after Asch, UK psychologists Steven Perrin and Christopher Spencer repeated his study, using engineering students as their subjects. This time, in almost 400 trials, only one went along with the majority view.[16]

The likely explanation for this shift from conformity is that Asch conducted his research in the post-war era at a time when conformity was the social norm. It was not until the end of the 1960s, with the rise of the counterculture movement which actively rejected traditional values and embraced individuality, that conforming to the view of the majority become a less dominant norm.

People who resist change, persuasion and contextual influence are less susceptible to manipulations than those whose attitudes are unpredictable, malleable and formed from moment to moment.[17] But at the same time, if people with strong attitudes accept a manipulation, they are more likely to resist subsequent attempts to make them acknowledge choice blindness.

Two powerful and related emotions that trigger choice blindness are disgust and moral outrage. In 1860, biologist Charles Darwin suggested disgust is an innate involuntary sensation that evolved to prevent our ancestors from eating rotten food that might kill them. Those most easily disgusted survived to pass on their genes, while the more nutritionally daring perished before reproducing. But disgust can be triggered by far more than unhealthy food. Social psychologist Jonathan Haidt illustrated the part played by disgust and its unconscious influences on our choices with a hypothetical story about brother and sister Mark and Julie.

The story went that while on holiday, Mark and Julie decided it would be fun and exciting to have sex. Haidt makes it clear that Julie was taking birth control pills, and Mark wore a condom. Both vowed never to repeat the experience and to keep what happened a secret. One that made them feel even closer to each other.

Mark and Julie would most likely have kept their tryst secret in real life, but since this was only a philosophical experiment, it was thrown open to the world to pass judgement, which they did in great numbers.

While most people interviewed were certain that what Julie and Mark did was wrong, they struggled to rationalise their disquiet. Some cited the risk of inbreeding, and most believed it either would or could cause them, or those close to them, emotional or psychological damage. Yet, as Haidt points out: 'The scenario was constructed in such a way as to rule out these possibilities.'

Having exhausted their objective reasons for their reaction, most admitted: 'I can't explain it; I just know it's wrong.' 'The subjects went on to exhibit all the trademark signs of a morally dumbfounded state,' Haidt reports, 'including confusion, a tendency to withdraw reasons, and the declaration of dumbfounding... their credulity regarding the non-occurrence of certain types of harm, but not their level of physical disgust, strongly and uniquely predicted their disapproval of the act.'[18]

Interested by these findings, Elizabeth Horberg and her colleagues conducted experiments to discover whether disgust is more likely to trigger moral outrage than other negative emotions, such as sadness. Their subjects watched a clip from a film intended to provoke either disgust or sadness.[19] To provoke disgust, the study showed an excerpt from the 1996 film *Trainspotting* in which the lead character plunges his hand into a faeces-smothered toilet. To elicit sadness, it showed a scene from the 1979 film *The Champ* in which a young boy witnesses his father's death. They then completed a moral judgement task in which they judged moral violations (such as keeping an untidy and dirty living space) and virtues (such as maintaining a healthy body) on a scale of 1–7 depending on how 'good' they believed each behaviour to be. The study showed those who were disgusted were more likely to pass moral judgements than those who were sad, which they argued is because disgust is more associated with contamination or purity.

You can test your own 'disgust' tolerance with this simple experiment by psychiatrist R. D. Laing. He proposed filling a tumbler with water and taking a sip. Spit the water back into the glass and take a second sip. Continue in this way until you can no longer bear to drink from the glass. The idea of putting more of the saliva-infused liquid into their mouth revolts most people, and most abandon the test after just three sips.[20]

But why? The water contains nothing that can harm you or did not originate in your body only a few moments before. In this, as in other life experiences, the point at which disgust takes over depends on your experiences and expectations.

Many of our choices are based on the same vague feelings that an action is right or wrong, disgusting or acceptable, without really being able to explain why. Many who enjoy eating shrimp express disgust at consuming other arthropods, such as crickets. And yet insects are a protein-rich, low-fat dietary staple in some parts of the world.[21] Which raises a more general question: to what extent do upbringing and experiences influence our choices?

Confabulations

As we can see from the above, many choices are made without people being consciously aware of why they are making that choice. But if people often make choices without knowing why, why don't they say so?

The likely answer is that they unconsciously invent reasons for those choices after the fact and then become convinced of their truth; these are termed pseudo-explanations or confabulations.

Confabulations, from the Latin *fabula*, for 'story', are sometimes called 'honest lies' since those uttering them genuinely believe they are speaking the truth. Even when, as in the case of the swapped photos or manipulated jam-tasting studies, people offer reasons to justify a choice they never made.[22]

IT'S YOUR CHOICE – OR IS IT?

Today, psychologists use the word to describe people who unintentionally make false reports about their actions, intentions, emotions and perceptions. Rather than admitting they don't know, the person comes up with wrong, if plausible, reasons for what they said or did.[23] When they repeat an 'honest lie' often enough, people firmly believe it is true.[24]

Indeed, people are more likely to confabulate when *presented* with false information than when they self-generate falsehoods.[25] If a witness incorrectly chooses someone in a police line-up, they will be far more confident if the person in charge tells them they are right.[26] This effect of confirmatory feedback appears to last over time, as witnesses will even remember the confabulated information months later.

'When giving answers to questions about our reasons for action, we do not introspect on the mental states that caused our action but rather come up with a story that makes it plausible why the action we performed is a reasonable response to the situation we faced,' suggest Derek Strijbos and Leon de Bruin. 'Cases of confabulation are understood as failed attempts at self-interpretation.'[27] Choice blindness and the often complex justifications we make when explaining a choice we never made highlight our inability to reason about our past actions.[28]

When the philosopher Jean-Paul Sartre announced, 'I am my choices. I cannot choose', he was more correct than he may have realised. Consider, for example, seeing a cup of steaming-hot coffee on your breakfast table.

How do you know it's a cup of coffee?

This may seem like a stupid question, but our ease of recognition disguises complex cognitive choices – rapid and effortless options of which we are never consciously aware.

'In a visual search task, a target stimulus must be discriminated from an array of distractor stimuli,' says neuroscientist Jeffrey Schall. 'Visual search for a single target among distinct alternatives, known as "feature search", requires a choice that can be based entirely on sensory processing. Search is more efficient if the target is conspicuously different from distractors – for example, a different colour or shape. Search is less efficient if the target is less discriminable from distractors – for example, a small difference in colour, shape, or shared features in a search for a conjunction of features.'[29]

Different parts of the brain process visual information regarding the shape of the cup and saucer, the colour and texture of the liquid, and information regarding its aroma and temperature from the olfactory organs before we can choose precisely what we see, smell, taste and touch.

Neuroscientists debate how all these choices are combined to create the perception of a cup of coffee. This is known as the 'binding problem'. The brain makes lightning-fast choices between different options before 'choosing' to perceive incoming sensory information as indicating a cup of coffee.

Yet these perceptions based on expectations can have wider impacts than simply taking a sip of tea that we believed to be weak coffee. On a summer evening, a colleague, her partner and two young children were driving to their holiday destination. They had set off early in the evening, hoping to avoid the rush hour. They were wrong. Traffic was heavy, and progress was slow. The children, who had missed their tea, were becoming hungry and irritable. While their mother wanted to stop for refreshments, her partner insisted they press on. As time passed, their chances of finding an open eatery faded.

While passing a lay-by, she suddenly spotted the message 'Hot Pies and Peas' emblazoned on the back of a large van. Eager to feed the children, she insisted on turning the car around and chasing after it. When catching up with the van, some twenty minutes later, she realised the sign read not 'Hot Pies and Peas' but 'Hoot Please and Pass'. Her expectation that it was a fast-food van had blinded her to what she saw.

In the next chapter, we describe how expectation blind spots prevent us from seeing what is there to be seen. Sometimes with severe consequences.

CHAPTER FOUR
Expectation Blind Spots

One spring morning in 1900, psychologist Norman Triplett, aided by a professional conjuror, baffled schoolchildren by making a ball disappear before their eyes. In a paper entitled 'The Psychology of Conjuring Deceptions', he explained how the magician, seated behind the teacher's desk, 'threw the ball about three feet in the air, catching it and letting the hands sink low behind the table. The second throw was four or five feet in height. On its return it was dropped between the magician's legs but his hands went up with the regular throwing movement and were held as if awaiting the descent of the ball.'

It never came down. So far as the children were concerned, the ball had vanished into thin air.

Of the 165 children shown the trick, 40 per cent of the boys and 60 per cent of the girls were deceived. One 14-year-old described how it was 'about one yard from the ceiling before disappearing', while another claimed to have seen it rise 'halfway up to the ceiling' before vanishing.[1]

Suggestions as to how this trick was performed ranged from improbable to impossible. Some youngsters claimed it must have been an inflatable ball that burst, and others that it had fallen behind a picture on the wall. One even believed she had spotted a black-costumed assistant, balanced on a ladder, who caught and hid it!

Magicians typically encourage such improbable pseudo-explanations to conceal sometimes simple, even banal explanations. Triplett's trick is a case in point. On his third throw, he had mimed tossing and preparing to catch the ball. Those claiming to have seen the ball fly upwards did so only because they expected to see it. Triplett became one of the first psychologists to demonstrate that we see what we expect to see rather than always seeing what is there. 'People's perceptions can involve predictions driven by long-term knowledge as well as perceptual inputs from the immediate past,' explain psychologists Gustav Kuhn and Ronald Rensink.[2]

Today's magicians employ these expectation blind spots in many tricks. For example, in the vanishing coin illusion they make a coin 'disappear' while throwing it from one hand to another. Similarly to Triplett's experience, spectators often report witnessing a coin leave the magician's hand, even though no coin was thrown to the other hand. Convinced it has passed from the magician's right hand to his left, spectators are baffled when he reveals his left hand is empty. People tend to see what they expect to see, which is reinforced by social cues like the magician looking at the hand to which the coin has been 'thrown' and jolting the hand of one side, mimicking what would have happened if the coin had really been thrown across. Such is the strength of expectation blind spots that you are likely to be deceived even when you know how the trick is done.

To watch this trick performed and discover how it is done, go to lewisandleyser.com/vanishing-coin or scan this QR code.

How Expectations Influence Perceptions

'If you can't quite tell what something is, but from your prior experience, you have some expectation of what it ought to be', says MIT director of education Mehrdad Jazayeri, 'then

EXPECTATION BLIND SPOTS

you will use that information to guide your judgment. We do this all the time.'[3]

Expectations arise from regularities in the world around us. For example, if our post always arrives before midday, we expect it to arrive then. Similarly, if we throw a ball into the air, we expect it to fall back to Earth.

Because horizontal and vertical lines occur more frequently than angled ones, we expect buildings, lampposts and telegraph poles to appear this way. Since light usually comes from above, we perceive circles with shadows at the bottom to be convex and those at the top concave, as shown below.[4]

Figure 12

Imagine walking down an unfamiliar street on a foggy day with everything blurry and indistinct, as in the picture on the next page.

As you struggle to identify and avoid potential obstacles, your eyes are endlessly bombarded by visual signals open to numerous interpretations. There are two main objects visible here and, while nothing identifies either of them, knowing you are on a city street where you expect to see pedestrians and vehicles, that's how you will most likely interpret these shapes.

The more stable the environment, the more reliable our expectations will be. The more chaotic, the less trustworthy.[5]

Figure 13

Where ambiguities occur, it is up to the observer to identify and resolve any perceptual biases. As perception researchers Aude Oliva and Antonio Torralba explain: 'The visual system makes assumptions regarding object identities according to their size and location in the scene.'[6] Expectations based on past experiences therefore help you make sense of what you see.

Expectations lead to predictions, assumptions and the acquisition of beliefs. These enable us to survive in a challenging and changing environment while only putting limited demands on our brain's processing capacity. Also called 'prior beliefs', expectations help us make sense of what we are presently perceiving based on similar experiences in the past. The shadow on a patient's X-ray image, which a less experienced intern could easily miss, jumps out at a seasoned physician. The physician's prior experience helps her arrive at the most probable interpretation of a weak signal.

By drawing on past experiences, we defy the arrow of time by recalling the past and travelling into the future through expectations of what will happen tomorrow, next week or next year.[7] 'We alone seem to have the capacity to create detailed imaginings of future possibilities,' comment psychologists Bradley McAuliff and Brian Bornstein. 'To create vivid simulations of non-existent places and situations, often involving many people, and, through effort and invention, transform those imaginings into realities.'[8]

American psychologist William James described the visual world of children as a 'blooming, buzzing confusion'.[9] The same can be said of perceptions. The certainty of our visual experience contrasts starkly with the ambiguity of images seen by the eye. Similar objects, such as the two bicycles below, may give rise to very different images on the retina while being experienced as nearly identical. At the same time, objects that appear identical can produce different visual experiences.[10]

Figure 14 *Figure 15*

Imagine looking at a picture of your grandmother. While the image on your retina might look like your grandmother, it could, says cognitive scientist Jacob Feldman, also resemble 'an infinity of other arrangements of matter, each having a different combination of surface and colour properties that happen to look like your grandmother. Your brain does not consider these far-fetched alternatives but rapidly converges on the most likely solution which determines what we consciously perceive.'[11]

When two people look at the same object, each assumes they see the same thing. In 2015, an internet phenomenon known as 'the dress' showed this to be false. Some claimed a photo showed a blue and black dress, while others were certain it was white and gold.

The debate went viral. Scientists said that some people inferred that the dress was in direct light and mentally subtracted yellow from the image to see the dress as blue with black stripes correctly. Others assumed the dress was in shadow, where bluish light dominates, and subconsciously subtracted blue from the image to see it as white and gold.

Because we fail to realise how much our judgements depend on implicit assumptions, we can be surprised to find others fundamentally disagree with our perceptual judgements.

'The interpretation occurs automatically when looking at a photo and in general when looking at a realistic scene or anything else in the natural environment,' explains psychologist Christoph Witzel. 'Our brain does not see but constructs reality. It infers an outside world from ambiguous input … the phenomenon of the dress nicely illustrates the power of unconscious assumptions and beliefs in perception in general.'[12]

So do you see the world as most others see it?

The most likely answer, in general terms, is almost certainly 'Yes, you do', but in many subtle and sometimes vitally essential ways, you do not. Nor does anyone else, not even those closest to you.

A mid-grey region on the retina might be a bright white object in dim light, a dark object in bright light, or anything in between. An elliptical shape on the retina might be an elliptical object face-on, a circular object slanted back, or anything in between. A shape on the retina might be a large object that is far away, or a smaller one nearby, or anything in between.

There is an infinity of possible scenes in everything we see. The central task of the visual system is to quickly and reliably decide among these alternatives. Visual science aims to discover the rules, principles or mechanisms the brain uses to do so.[13]

EXPECTATION BLIND SPOTS

The 19th-century polymath Hermann von Helmholtz was one of the first scientists to appreciate that the brain must recover the three-dimensional world that gives rise to a two-dimensional image on the retina, a process he described as 'unconscious inference'.[14] Following Helmholtz, scientists now regard human vision as a statistical inference engine whose function is to infer the probable causes of sensory input. This is where the work of 18th-century clergyman, amateur mathematician and author Thomas Bayes enters the story. After the clergyman's death, a friend named Richard Price went through his unpublished papers at his widow's request. Among them, he discovered Bayes' writings on probability and recognised their importance in making and refining predictions.

Bayesian inference can be understood by rolling a die and determining whether it is fair or unfair based on the results you roll.

Before rolling the die, you have some belief about whether the die is fair. Let's say you start with a 50 per cent belief that the die is fair (meaning all sides have an equal chance of 1/6), and a 50 per cent belief that it is unfair (for example, it might be biased to roll a six more often).

You roll the die ten times and get a six on four occasions.

Based on this data, you can use Bayes' theorem to update your belief about whether the die is fair or biased based on the *likelihood* of getting the data (rolling four sixes) you did. If the die had been fair, the chance of getting four sixes in ten rolls is low, but if the die was biased towards the six, the chance of getting four sixes is higher.

These simple facts form the basis of Bayesian inference: a statistical procedure for combining new data with prior information to improve the accuracy of a statement about the likelihood or probability of an event. The result is called a posterior probability.

Your visual system is widely believed to function in a similar way. It guesses what it is seeing and refines that guess by seeking additional information to confirm or disprove its initial prediction or hypothesis.

For instance, a widespread unconscious prediction is that if something moves, it must be alive. This arises through Bayesian inference. The power of this expectation is illustrated by audience reactions to magic performed by a robot dog in one of Keelan's stage shows.

Although it does not look anything like its flesh-and-blood equivalent, the robot moves in the same way as a living creature. Audiences applaud and cheer when it performs a trick but express outrage at Keelan's 'cruelty' when he is seen kicking it on social media.

To see how a television audience reacts to Keelan's robot dog when it performs a fantastic card trick, go to lewisandleyser.com/robot-dog or scan the QR code on the next page.

Psychologists have long realised that while we are aware of many of the associations we make, others are buried so deeply in our subconscious that we are unaware of them and how they influence our behaviour.

Figure 16

Implicit associations, or stereotypes, are the frameworks individuals construct to organise and make sense of the vast amount of information in the world. Think of them as the mental blueprints from which our expectations are constructed. They guide our perceptions and influence how we interpret and respond to new experiences.

From our earliest years, and possibly while we are still in the womb, our brains make pairings between thoughts and feelings, actions and attitudes. As we grow up, we acquire associations for our society and culture from those around us through what is known as the 'halo effect', sometimes described as the 'what is beautiful is also good' principle.[15]

In the 1920s, psychologist Edward Thorndike asked military officers to list about a dozen others of the same rank and rate them on their physique, bearing, neatness, voice, energy and endurance.[16] He aimed to determine how ratings of one attribute, such as having a fine physique, correlated with their assessments of other attributes, such as intelligence or leadership.

Thorndike reported that a high rating on one attribute was often correlated with a high rating on another. Equally, negative ratings on one attribute were correlated with negative ratings on another. He discovered, for example, a closer association between an officer's physique and his leadership skills than between his physique and character.

An ability to make rapid judgements about those we have just met in terms of beliefs about people we already know is one of the most valuable tools in our social perception kit. Implicit associations have been dubbed the 'sluggard's best friend'[17] because they speed up decision-making, saving time and mental energy.

As Gordon Allport pointed out in his 1954 book *The Nature of Prejudice*, 'We like to solve problems easily. We can do so best if we can fit them rapidly into a satisfactory category and use this category as a means of prejudging the solution. We tend to do so as long as we can get away with coarse overgeneralisations ... Why? Well, it takes less effort, and effort, except in the area of our most intense interests, is disagreeable.'[18]

On the downside, as we explained when describing Norman Triplett's magic trick, these associations encourage us to see what we expect rather than what is there to see, leading to rushed judgements based on stereotypical assumptions. Beautiful people are generally seen as more intelligent and honest than those deemed less attractive, for instance,[19] and one study even found jurors were less likely to believe that attractive people were guilty of criminal behaviour.[20]

While expectations are often accurate, when erroneous they can and have ended in tragedy, which is especially true when affected by stereotype bias.[21]

When 47-year-old Donnie Sanders was pulled over for a minor traffic violation, his encounter with the Kansas police ended not with a ticket but his death. Believing he was holding a weapon, the officer fatally shot him. He was later found to be unarmed and only had a cell phone. Because a Kansas police officer expected an African American to be carrying a gun, that is what he 'saw'.[22]

Prejudices like this are far from rare in the US. Around midnight on a winter's night in 1999, Amadou Diallo, a 22-year-old West African immigrant, was standing in the doorway of his apartment building in the Bronx, a borough of New York. Four plain-clothes police officers searching for a rape suspect saw him and believed he resembled the suspect they were tracking. When ordered to remain motionless, Diallo reached into his trouser pocket. Believing he had a gun, the police fired a total of 41 shots, 19 of which hit and killed the young man. Even though he was unarmed, all four officers were later acquitted of any wrongdoing in the case.

'The police could not have known for certain that Diallo was harmless,' comments psychologist Joshua Correll. 'In the dark, they had ordered a potentially dangerous man to freeze, and that man reached for something.'[23]

Their decision to open fire would never have been questioned had Diallo been armed. But their decision to shoot him raises a fundamental question equally relevant in the case of Donnie Sanders. Would the police have responded differently if Diallo had been white?

Research evidence suggests they would. A study of police decision-making, by Geoffrey Alpert and his colleagues, found that most people stopped by police were young males (74 per cent) and minority group members (71 per cent), with Black individuals more likely to arouse suspicion and get stopped than white individuals (71 per cent vs 31 per cent).[24]

In another study, Joshua Correll and his colleagues showed US officers a video depicting a young man removing something from his pocket. The officers were asked to respond instantly if they expected the suspect to be pulling out a weapon. When he was Black, eight out of ten officers concluded he was drawing a gun when, as in Sanders' case, he was only taking out his phone. The researchers dubbed this mistaken expectation 'shooter bias' and found it was present irrespective of the officer's own ethnicity.

'These studies have demonstrated that a target person's ethnicity may influence the decision to shoot,' says Correll. 'Participants showed a bias to shoot African American targets more rapidly and/or more frequently than white targets. The implications of this bias are clear and disturbing. Even more problematic is the suggestion that mere knowledge of the cultural stereotype, which depicts African Americans as violent, may produce Shooter Bias and that even African Americans demonstrate the bias'.[25]

Expectations based on stereotypes, while speeding decision-making, also reduce the accuracy and reliability of our perceptions. While stereotypic beliefs about the role of women in society may 'enable one to assume a woman in a dark room is threading a needle rather than tying a fishing lure', say psychologists Daniel Gibson and Gregory Hixon, 'they may also cause one to mistakenly assume that her goal is embroidery rather than cardiac surgery'. Although stereotypes are psychologically fundamental, they may also be socially pernicious, and psychologists have long searched for ways to resolve this.[26]

Accurate expectations about others can only be achieved by abandoning fast but lazy stereotypic thinking and accepting people as the unique individuals they are.

However, while some expectations are explicit and can be readily expressed in words – 'I expect to see you next Monday', you might tell a friend – others are implicit. These implicit expectations guide our long-term behaviour and play a vital role in everyday experiences like driving a car, organising a meeting or dining at a restaurant, without our conscious awareness of their existence.

Implicit association maps, then, are the mental frameworks individuals construct to organise and make sense of the vast amount of information in the world. These internal maps guide perception, influencing how people interpret and respond to new experiences. Think of association maps as cognitive blueprints that shape our expectations and understanding of relationships. If, for example, someone has a positive association with dogs due to fond childhood memories, their internal map may include positive emotions linked to the idea of dogs. Conversely, if someone had a negative experience with public speaking, their association map for public speaking may evoke anxiety and discomfort.

However, sometimes, as in the case of racial stereotypes, people are not consciously aware of their implicit associations. One way of still being able to study these associations is to use the Implicit Association Test (IAT) developed by Anthony Greenwald and Linda Krieger.[27] During the IAT participants are asked to categorise words or images positively or negatively, for example as 'good' or 'bad', 'friendly' or 'hostile', and the time taken to respond is carefully measured. The faster participants sort a word or image into a positive or negative category, the stronger they must be linked in that person's mind.

Imagine sorting a deck of playing cards in two different ways: first by placing hearts and diamonds (both red symbols) into one pile and clubs and spades (both black) into another, and then by sorting clubs and hearts into one pile and diamonds and spades into the other. Because the first task allows you to take advantage of a common feature – the different colours of the two groups – the task will be quicker and easier than the second task, where you have to pay attention to the symbols rather than just the colours.[28]

Overcoming Expectation Blind Spots

However, as most people are unable to take IAT tests to discover whether they hold implicit associations, becoming more aware of such blind spots in everyday life requires intentional self-reflection and an openness to challenging one's assumptions. One effective strategy is practising mindfulness techniques, such as meditation or mindful breathing, which can heighten self-awareness and help individuals observe their thoughts without instantly judging them. This approach makes it easier to identify biases influencing one's perspective.

Additionally, seeking diverse perspectives is crucial. Exposing oneself to various viewpoints, opinions and experiences through reading diverse sources and conversing with people from different backgrounds can broaden one's worldview. Regularly questioning assumptions and beliefs is another critical practice. This involves asking why particular views are held and whether they are based on evidence or on preconceived notions. Establishing a habit of self-enquiry can reveal hidden biases.

Reflecting on decision-making processes is also essential. Individuals can analyse their choices to determine whether biases have influenced their decisions or if alternative viewpoints have been adequately considered. Staying informed about cognitive biases and common conceptual blind spots provides a foundational understanding, enabling individuals to recognise and address these psychological phenomena in their thinking.

Seeking feedback from others is also a valuable strategy. Creating an open environment where constructive feedback is encouraged allows friends, colleagues or mentors to provide insights into blind spots that may not be apparent to the individual. Regularly updating knowledge on various topics helps prevent outdated or inaccurate beliefs from influencing perspectives.

Challenging stereotypes is a proactive step in addressing biases. Individuals can identify and question stereotypes, consciously treating people as unique entities rather than conforming to preconceived notions. Reflecting on emotional responses in different situations is also illuminating. Emotions can indicate underlying biases, and recognising strong emotional reactions prompts examination of their root causes.

Expectation blind spots, however, don't only impact how we see other people, but can have major consequences for other areas of our lives. When jetting off on holiday or on a business trip, the last thing you expect is for the pilot to land on the wrong runway, let alone at the wrong airport. But that happened in 2006 when the 39 passengers aboard a Ryanair flight from Liverpool to Derry found themselves not at the city's airport but at a military base 8km away. Such a mistake is far from unusual, having occurred in the US alone

150 times in the past 20 years. Even more frequent are occasions when pilots mistakenly landed on a taxiway parallel to the intended runway. This has happened almost 300 times.[29]

But why should an experienced flight crew, who have often landed at the correct airport on many occasions in the recent past, make such an obvious mistake? According to Linfeng Jin and Edmund Lo, one of the main reasons is a conflict between what they see through the flight deck windows and a mental map of the airport they hold in their heads.[30]

Similarly, while everyone likes to believe they will be in safe hands when visiting a doctor or spending time in a hospital, evidence suggests this may not always be true. Blind spot errors in diagnosis and treatment in the US alone were responsible for more than a million excess injuries, leading to some 100,000 avoidable deaths and costing medical authorities alone over $20 billion in lawsuits.[31] Here are two tragic cases reported by the Institute for Safe Medication.[32]

A nurse administered the blood-thinning drug heparin. She expected the label to read ten units per millilitre, and consequently this is what she 'saw', rather than the stated 10,000 units. The patient died.

A doctor picked out a syringe and, expecting it to be filled with morphine, administered the drug. Rather than morphine, it was hydromorphone, an opioid-based painkiller five times stronger. The patient died.

'All of these real-life errors, and many more in health care and other industries, have happened under similar circumstances: the person performing the task fails to see what should have been visible, and later, they cannot explain the lapse,' says medical researcher Matthew Grissinger. 'These types of accidents are common and can be made by intelligent, vigilant, and attentive people.'[33]

As we will explain later, expectation blindness is exacerbated by stress, boredom, powerful emotions such as anger and fear, and exhaustion.[34] When expectations are violated, our reactions will depend on the number of previous confirmations and disconfirmations experienced. Where an expectation has been confirmed numerous times, disconfirmations are less likely to lead to a change in opinion.

If, for example, someone has always behaved aggressively or dishonestly towards you, robust evidence may be needed before you change your view about them. You might not even notice if they perform an act of kindness and generosity. Or, if you do, you might dismiss it as insignificant. That individual may have to behave that way on numerous occasions before you change your expectations of them.

'One-time, mild expectation-discrepant events are easier to ignore than more frequent, more discrepant events,' say Martin Pietzsch and Martin Pinquart.[35] The more significant the discrepancies between expectations and disconfirming events, the more you trust the credibility of expectation-disconfirming information, and thus the more profound the change in expectations. Unambiguous expectation-disconfirming details may also be more likely to lead to expectation change than ambiguous information because the latter would be more challenging to process.

Expectations are also less likely to change when those we trust have the same expectations. Research has shown, for example, that regular encounters with drinkers voicing positive expectations about consuming alcohol lead to similarly positive expectations among non-drinkers.[36] This is one reason why the 'echo chamber' confirmation and support many social media sites provide is so comforting.

Overconfidence also increases our chances of experiencing an expectation blind spot. It can cause us to misinterpret the information supporting that expectation or downplay anything challenging. Sometimes with tragic results.

In July 2021, President Joe Biden assured Americans there were no circumstances under which they would see 'people being lifted off the roof of an embassy of the United States in Afghanistan'. Five weeks later, they were seeing just that. While the reasons for the catastrophe that followed America's decision to pull all its troops out of Afghanistan are many and complex, experts agree that expectation blindness played a significant role. As one intelligence insider put it: 'Officials were blind to the truth of what was happening on the ground.'[37]

No matter how optimistic you are or how insignificant it is in the general scheme of things, the immediate response to the failure of a positive expectation is usually disappointment, anger, frustration and denial. When passed over for a rival, an employee expecting a well-deserved promotion usually experiences all these emotions immediately after being given the bad news.

We must examine the two brain regions illustrated below to understand why they are primarily responsible for processing information concerning that failure.

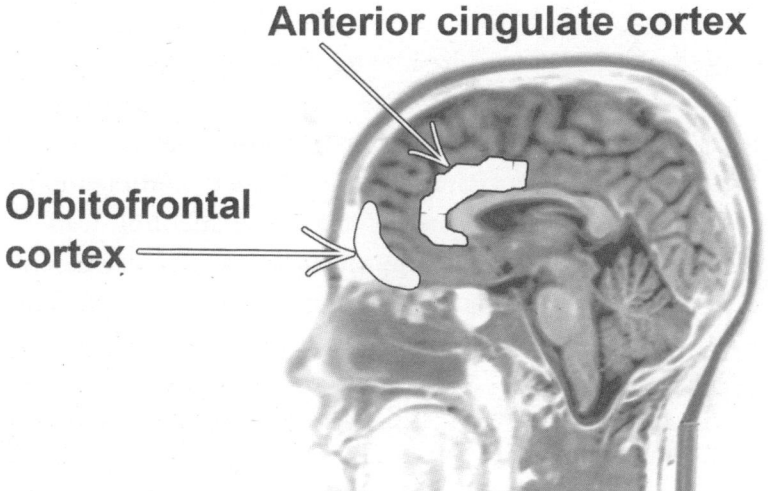

Figure 17

Imaging studies have detected heightened activity in the anterior cingulate cortex (ACC) when expectancies are disconfirmed. Crucially, this region constantly monitors and detects discrepancies between what was expected and what was delivered,[38] and is strongly linked to the brain's two major centres of emotion, the thalamus and the amygdala, which have powerful effects on emotional and cognitive processes. The posterior orbitofrontal cortex, a region with direct neuronal connections to the anterior cingulate, also detects the failure of expectations.[39] This suggests that if a vital expectation is violated – you had expected your partner to be faithful, your boss to award a promotion, or your child to follow the career you desired – the pain and fear experienced is similar to being physically assaulted.[40]

So far in this chapter, we have only considered the consequences of negative expectations. To conclude, let's look at a study that, initially at least, seemed to demonstrate that optimistic expectations can produce expectation blind spots resulting in *positive* outcomes. During the 1960s, psychologist Robert Rosenthal and school administrator Lenore Jacobson administered an IQ test to Californian elementary school children. The children's teachers were then given the names of a few pupils described by the researchers as 'intellectual bloomers': youngsters destined to outperform their classmates and achieve higher-than-average grades.[41]

The pupils had been randomly selected, and yet months later it was found that these youngsters had performed significantly better than the rest of their class. Teacher expectations of success had enhanced the children's classroom attainment. When they believed a child would do well, teachers only saw what they expected to see, their accomplishments, while remaining blind to their failures.

Among the case histories was that of José, a Mexican-American pupil who, in a year, moved from being one of those at the bottom of the class to above average. Another child, Maria, went from being a 'slow learner' to a 'gifted child', with one reviewer commenting that 'the implications of these results will upset many school people, yet these are hard facts.'[42]

In the decade in which it was published, Rosenthal and Jacobson's book *Pygmalion in the Classroom* attracted more mass media attention than any other psychological study, striking a responsive chord among millions seeking an explanation of the educational problems of children from low-income areas.

Based on extensive research, it has been found teachers' expectations impact the students' willingness to learn, how well they get along with the teacher and their fellow students, as well as how they feel about themselves.

As to whether these expectations actually boost IQ, however, educational specialists and statisticians are far more dubious.[43]

A positive effect of Rosenthal and Jacobson's study was the attention paid to how teachers' expectations significantly affect classroom performance. At the same time, it is evident that the Pygmalion effect, if any such thing exists, is an extremely subtle and elusive phenomenon that acts through teachers without their conscious awareness.

Expectation blind spots, then, are a pervasive and powerful force in our daily lives. However, as will by now be apparent, they are just one of the many ways our eyes and brains can be misled. Take, for example, the picture on the next page.

Does this depict a girl sitting on a giant saucer?

Or is it an illusion?

Your answer will depend on how your eyes and brain interpret the image. The extent to which they are fooled is the topic of our next chapter – illusional blindness.

Figure 18

CHAPTER FIVE
You Could Have Fooled Me!

Which of these tables is longer?

Figure 19

If you said the one on the left, an answer offered by 95 per cent of those seeing this illusion, you are wrong. They are the same size. Should you doubt this, go to lewisandleyser.com/shepards-table or scan the QR code.

'Visual illusions, ambiguous figures, and pictures of impossible objects are inherently fascinating,' says artist Roger Shepard, who created the illusion we have modified above. 'The world that we have relied on for solidity and stability shudders and shifts unpredictably, as if in a dream.'[1]

At the end of the 19th century, psychologist Oswald Külpe expressed the intellectual climate of the era when he wrote that perceptual illusions like the one shown in Figure 19 are 'subjective perversions of the contents of objective perception'.[2]

Today, all this has changed. The systematic study of illusions has provided vital clues about visual architecture and vision processes and exposed the extent to which everything we see or think we see is an illusion.

But what exactly is meant by an 'illusion'?

According to psychologist Richard Gregory, there are two types: those with a physical cause, such as a stick partly submerged in water appearing to bend, and knowledge-based, cognitive illusions 'due to misapplied knowledge employed by the brain to interpret or read sensory signals'.[3] It is with the latter type of illusion that we are concerned here.

Look at Figure 20. This partly blocked-out photograph of a couple jumping on a beach may appear naked. But it's your brain, not your eyes, that creates an impression of their nudity.

Figure 20

Figure 21

Once the bubble mask is removed, both people are seen to be wearing bathing costumes.

'The pornographer lurking in your brain has been especially aware of human nudity since your birth,' explains researcher Kyle Hill. 'A likely outcropping of evolutionary pressure on reproduction ... he is familiar with nudity, but not with swimsuits.'[4]

Unrestrained by social convention, Bayesian probability means we assume that a continuation of bare skin is the most probable scenario. With a choice between the ease of seeing nudity and the complexity of conjuring the wide variety of possible clothing, the brain sees the people in the image as naked.

As explained in the previous chapter, we are confident that 'seeing is believing' and that the 'evidence of our own eyes' is our most convincing proof of objective reality. 'If an eyewitness has "seen it with the naked eye",' comments psychologist Claus-Christian Carbon, 'judges, jury members and attendees take the reports of these precepts not only as strong evidence but usually as fact despite the active and biasing processes based on perception and memory. Indeed, it seems that there is no better, no more "proof" of something being factual knowledge than having perceived it'.[5]

Our belief that what we see is all there is to see is so strong that we only doubt it when visibility is poor, someone's vision is faulty, or the eyewitness is under great stress or mentally incapacitated.

But the visual system does far more than mirror the outside world, and our visual perceptions go far beyond information provided solely by the eyes.

Much of what we think we see lacks any direct physical counterpart and so is called an illusion. Because illusions reflect the brain's role in perceptual organisation, they offer an excellent opportunity for linking what we see with the mental activity involved.

The partly occluded image in Figure 20 exemplifies a type of illusion known as an amodal illusion, one that experimental psychologist Vebjørn Ekroll describes as producing 'a curious feeling that the hidden parts are really there and that they have a definite shape, although they are obviously not experienced in quite the same way as directly visible objects.[6]

Many illusions we discuss in this chapter produce the same curious feelings. They also provide invaluable insights into how vision works and why what we should see remains unseen.

Modal and Amodal Illusions

When looking at objects partly hidden from view, we use our imagination – which philosopher Robert Briscoe describes as the 'mental glue ... that binds the visible parts of partially occluded objects together into discrete, coherent wholes'[7] – to 'fill in' the missing parts. If, for example, you saw a dog partially concealed by a bush, you would not conclude that there was only half a dog but would mentally complete the occluded portion.[8]

Figure 22

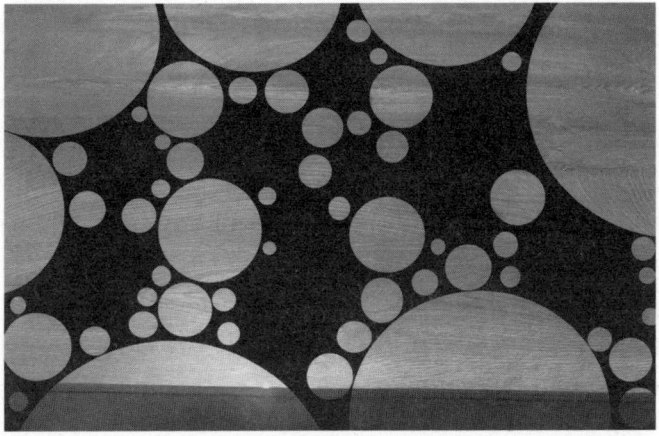

Figure 23

In the 'naked people' illusion, your brain 'fills in' those parts hidden from view – a process known as amodal perceptual completion, which links sensory and cognitive processes.

However, there are occasions when perceptual completion is far more difficult to achieve. Study Figure 22, which contains various objects, from a Rubik's Cube to a hamburger. Then, shut your eyes and try to imagine the scene with your 'inner eye'. You will probably find the task reasonably straightforward, with at least two or three items and their locations readily coming to mind.

Now repeat the process while looking at Figure 23, which is an occluded version of the same image. It will be far more difficult, if, indeed, you can do it at all.

Even when you know what is hidden, an occluder such as the bubble mask interferes with your ability to imagine what lies behind it. As Ekroll puts it: 'What you expect to be there is experienced as curiously "absent".'[9]

As well as amodal completion – the ability to see an object even when it is partly covered – there is also modal completion. This is the brain's ability to perceive a non-occluded object where there is confusing and disordered sensory information of some sort.

We effortlessly recognise the shape of black-footed penguins both in the photograph below and the modified image on the next page, which shows only their outlines.[10]

What's happening here may be better described as an example of 'perceptual completion' or 'contour completion' – a normal function of our visual processing system – whereby the brain completes the missing contours based on the available visual information and our prior knowledge of penguin shapes.

The Kanizsa triangle, depicted in Figure 26, was created by Italian psychologist and artist Gaetano Kanizsa in 1955, and is another example of an illusory contour.

Figure 24

Figure 25

The visual system creates an imaginary triangle whose interior, without any change in contrast, appears brighter than the background.[11]

Both examples help us understand the central question of visual perception: why do things look as they do?

In the natural world, modal perceptual completion may be the difference between life and death. Many birds hunt in conditions of low visibility. Raptors like falcons, ospreys and kestrels, for example, hunt in woods and forests, and owls often hunt at night – circumstances in which the contours of prey are unclear, absent or incomplete.[12]

Modal completion allows potential prey to minimise their contours using camouflage that blends with the surroundings. A chameleon, for example, can adopt the colour and texture of its environment.

Figure 26

Anamorphic Illusions

An even stranger, far older illusion than the amodal or modal types is the 'anamorphic' illusion – from the Greek *ana* ('again') and *morphe* ('shape' or 'form'). Here an image appears distorted from one angle and normal when seen from another or viewed in a mirror.

Figure 27 is a modern version of this illusion created for one of our studies where we asked which of the four objects is an illusion: the mug, playing cards, Rubik's Cube or marker pen.

Figure 27

From a front-on viewing position, it's impossible to tell. However, when seen from another angle, as in Figure 28, the answer immediately becomes apparent. It's the Rubik's Cube.

Figure 28

YOU COULD HAVE FOOLED ME!

To see this illusion in action, visit lewisandleyser.com/anamorphic or scan the QR code.

Anamorphic illusions, then, underline the importance of perspective in making sense of the world. Understanding what one is seeing can crucially depend on the viewing position taken.

Over the centuries, artists have used anamorphic illusions to create aesthetically pleasing effects in paintings or even or to hide secret messages within them. Roman historians Pliny and Tzetzes wrote about a competition between the sculptors Alcamenes and Phidias to create an image of Minerva, goddess of wisdom and patron of the arts. While Alcamenes' sculpture of the goddess was beautiful, that of Phidias was grotesquely disproportioned when viewed in his studio. Once mounted on pillars, however, the altered perspective rendered Phidias' Minerva beautiful too.

After Charles Edward Stuart, claimant to the English throne, was defeated at the Battle of Culloden in 1746, support for him became a treasonable offence, with anyone displaying his image risking execution. Followers got around this prohibition using anamorphic portraits, which could only be seen correctly if viewed at a certain angle. One anamorphic painting by an unknown artist appeared on a wooden tray. When a wine glass or metal cylinder was placed on the tray, a portrait of Charles emerged as a reflection and could be toasted by his supporters. Should a non-Jacobite call, however, the cylinder could be removed from the tray, hiding the host's support for the Stuart cause.

Figure 29

Hybrid Illusions

Another illusion popular with artists is that of the hybrid image that varies according to your distance from it. When looking at the picture below from close-up, you see a wolf, yet as you get further away, it turns into a cat.

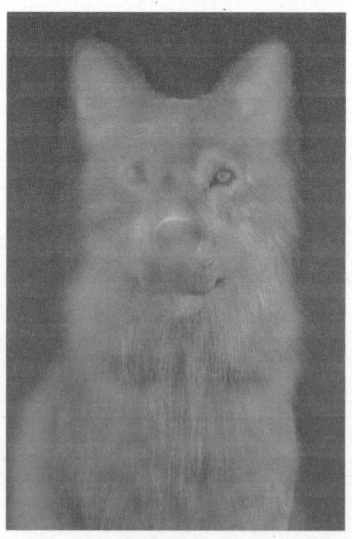

Figure 30

If you have difficulty making the switch, go to lewisandleyser.com/hybrid or scan the QR code to watch a video of three hybrid transformations. One shows the wolf transforming into a cat and vice versa, the second an old woman morphing into a young one, and the third the two authors of this book turning into each other.

In Figure 30 you recognise the wolf using fine-grained discrimination and the cat using coarser-grained frequencies. The difference lies in their spatial resolution, or how close two adjacent structures must be to be correctly resolved as separate structures rather than a single one. By changing your viewing distance, the brain automatically switches from one to the other. The global precedence hypothesis of image analysis ('not seeing the forest for the trees') implies a coarse-to-fine frequency analysis of an image, where the low spatial frequency components dominate early visual processing.[13]

Illumination Illusions

As we saw in the previous chapter's illustration of convex and concave shapes (Figure 12), the way something is lit plays a crucial role in how it is seen.[14]

The dinosaurs at the top of the image on the next page appear lighter grey, while those at the bottom look darker. Yet they are identical.

Known as the Anderson–Winawer lightness illusion after psychologist Barton Anderson and neuroscientist Jonathan

Figure 31

Winawer, this embedding of identical target patches in different surroundings is one of the techniques most widely used to investigate the effects of illumination on perception. It reveals a great deal about how our visual system functions.

In the illusion below, your brain adjusts the grey tones to compensate for what it perceives as light sources and shadows. Although hard to believe, square A is the same shade of grey as square B.

If you don't believe us, check our video at lewisandleyser.com/checkerboard or scan the QR code on the next page.

Figure 32

While several factors determine the light received by our eyes, such as the amount of light absorbed, reflected or deflected by the current atmospheric conditions (for example haze or other partially transparent media), only the proportion of the reflected light is of interest to the visual system.

'To accurately recover lightness, the visual system must somehow disentangle the contributions of surface reflectance from the illumination and atmospheric condition in which it is embedded,' say Barton Anderson and Jonathan Winawer, 'There is a growing body of data showing that the visual system can make systematic errors in estimating surface reflectance, the opacity of transparent surfaces or media and the amount of illumination striking a surface.'[15]

Autokinetic Illusions

On a clear night, take a moment to stare at the stars. Focus on a single one, and it will appear to move slowly across the heavens. Although the stars do gradually move in the night sky as the Earth turns, this effect is far too slow for us to discern and what you see is an illusion. German naturalist Alexander von Humboldt was the first to record seeing this from a Tenerife mountaintop in June 1799. He called it *Sternschwanken* (a made-up word with no direct translation, the closest being an 'unsteady star') and the term became a staple of astronomical literature for the next half-century. Not until 1857 did psychologist Kaspar Schweizer demonstrate that this movement occurs not in the sky but in the observer's brain.[16]

Vision researcher Richard Gregory believes this happens when the brain misinterprets eye movements caused by muscle fatigue as movements of the light.[17]

As you might expect, such autokinetic illusions, as they are termed, can have dangerous consequences. It was approaching midnight during the second Gulf War in 2003 when US Marines saw what they took to be numerous lights moving towards them from some 6km away. Believing they were an enemy column of some 140 vehicles, their nervous platoon commander requested urgent assistance. 'During the next few hours, attack jets dropped nearly 4.5 tonnes of bombs on the suspected position of the alleged enemy convoy,' reports Evan Wright, a journalist embedded with the Marines. 'It's a spectacular show ... we watch numerous smaller bombs flash and count two huge mushroom clouds roiling up in the night sky.'[18]

Yet when daylight came, it revealed scores of dead and injured civilians and bombed-out villages but no signs of enemy activity. The 'moving' lights the Marines had seen 6km away turned out to have been stationary lights from a town 17km away. The trigger-happy Marines had been taken in by an autokinetic illusion.

The same illusion can cause pilots flying at night to become disorientated. Aircrew can view a small, fixed light on the ground as moving and sometimes chase after it under the mistaken impression that it must be light from another aircraft. It can also be a problem for helicopter pilots as they attempt to land or hover at night when their only visual reference is a solitary light.

Not only has this resulted in fatal air accidents, but the illusion persists even if pilots are trained to recognise it. Statistics show that up to 80 per cent of all aviation accidents are attributable to human error. An accident rate of between 25 and 30 per cent has remained unchanged for decades, and current trends suggest a possible increase in such events.[19]

Pictures that Fool Your Eyes

For millennia, artists have created *trompe l'oeil* pictures that 'fool the eye' by creating 2D images that project on to the retina like a 3D scene. In the 4th century BCE, there was a competition between two of the city's most renowned artists, Parrhasios and Zeuxis, to see who could produce the most realistic *trompe l'oeil* painting. At first, Zeuxis claimed victory after painting a bunch of grapes so lifelike that birds swooped from the sky, hoping to eat them. But when Zeuxis inspected his rival's painting, he was annoyed to find it concealed behind a heavy curtain. He realised it was a painting only after he attempted to remove the drape. 'Zeuxis acknowledged his defeat,' says art historian Ernst Gombrich, 'because while he had tricked birds, the curtain of Parrhasius had deceived a man and fellow artist.'[20]

Figure 33, which we created for a study on social media, may help you judge the effectiveness of *trompe l'oeil* artworks created using modern technology.

Figure 33

Seventy-two per cent of those viewing this image believed it to be a genuine hole in the wall, with only 28 per cent recognising it as an illusion. With the rise in AI and digital technology, creating images and movies designed to deceive is becoming even easier.

While artists have been creating *trompe l'oeil* pictures for thousands of years, a more recent phenomenon has been the use of ambiguous images in psychological experiments. One of the first psychologists to do this in academic research was Edwin Garrigues Boring. In his 1930 paper entitled 'A New Ambiguous Figure'[21] he used a somewhat sexist cartoon created in 1915 by artist William Ely Hill for the American humorous magazine *Puck* captioned 'My Wife and My Mother-in-Law'.[22]

Whether you see a young or older woman when looking at the picture depends, to some extent, on how old you are. In a study involving hundreds of subjects, Michael Nicolls found

Figure 34

that the under-30s were most likely to describe a young woman facing away from them and looking over her right shoulder; they remained psychologically blind to the alternative. Those over 30 typically see an older woman with a large nose and protruding chin but fail to see the younger one.[23]

Confounding Our Inferences

Some eighty years before Hill drew his cartoon, minerologist Louis Albert Necker wrote a letter to the London and Edinburgh Philosophical Magazine and Journal of Science in which he described how a rectangular crystal would change shape as you look at it.[24]

When viewing the prism depicted in Figure 35 as having the N at the bottom right front and the M on the top left rear, you remain blind to the alternative in which the M appears in the top left front and the N at the bottom right rear.

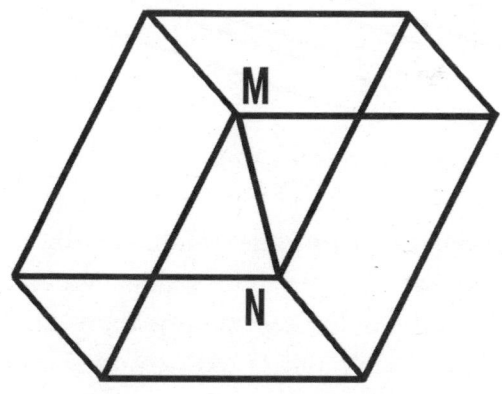

Figure 35

One of the insights offered by such illusions is that visual perception is not necessarily valid. Instead, what we see is based on deeply entrenched inferences acquired during our earliest months – on implicit, unspoken expectations operating outside conscious awareness and voluntary control.

From the start of postnatal life, infants are active perceivers and participants in their own visual development, showing consistent preferences for some stimuli over others: patterned rather than unpatterned objects, curves more than rectangles, moving versus stationary, three-dimensional not two-dimensional forms, and high- versus low-contrast patterns.[25]

'We do not first experience a two-dimensional image and then consciously calculate or infer the external three-dimensional scene that is most likely, given that image,' explains Ernst Gombrich. 'The first thing we experience is the three-dimensional world – as our visual system has already inferred it based on the two-dimensional input.'[26]

An even greater cause of optical confusion than the prism above is an ambiguous illusion known as 'the devil's tuning fork'.[27]

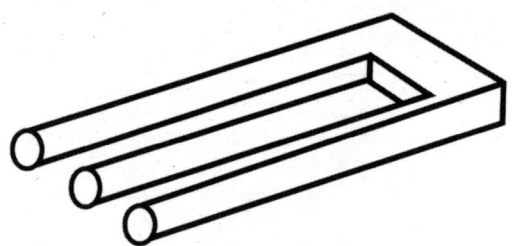

Figure 36

The middle prong appears to be simultaneously there and not there. As Richard Gregory points out: 'We can adopt no perceptual hypothesis for reconciling these features of the drawing into a possible object, and so no possible object – hypothesis is selected by the retinal image given by this drawing, and we cannot see it.'[28]

Size Illusions

In a study we conducted on how an object's apparent size determines what we think we see, Keelan was filmed walking down a line of parked cars.

Figure 37

See whether you can detect anything strange in this still from one of our research videos. For nine out of ten people, everything looks normal. If you are among them, you'll be surprised by how wrong your perceptions and expectations were after visiting lewisandleyser.com/car-park, scanning the QR code or going to the end of this chapter.

In 1911, psychologist Mario Ponzo demonstrated a fundamental principle of visual perception: when identical objects are placed between converging lines, the object closer to the point of convergence appears larger. This principle, now known as the Ponzo illusion, is powerfully demonstrated by railway tracks. As shown in Figure 38, two identical black lines are placed across the tracks. Although both lines are exactly the same length, the upper line appears longer because our brain interprets the converging railway tracks as parallel lines receding into the distance - just as they do in the real world. This unconscious depth processing causes us to perceive the 'more distant' line as longer.[29]

Figure 38

If you'd like to check this, take a look at the video we produced for one of our vision studies at lewisandleyser.com/tunnel

'Because we are generally unaware that we are imposing a perceptual interpretation of the stimulus,' says Richard Shepard, 'we are generally unaware that our experience has an illusionary aspect. The illusionary aspect may strike us only after we are informed, for example, that the sizes or shapes of lines or areas that appear unequal are, in fact, identical in the picture.'[30]

This is another illusion that can have some dangerous consequences. We often underestimate the time it will take an approaching object to reach us from the speed with which its size increases.

Known as the problem of optic flow, this illusion arises from the computational challenge of accurately determining an object's speed and direction based solely on changes in what the eyes see from one moment to the next.

'The illusion fools drivers into underestimating how long it will take them to stop,' explains the study's author, Doug Stewart. 'By discovering their mistake, it is too late to avoid an accident.[31]

This also prevents pedestrians from recognising the danger they're in of being struck by a speeding vehicle or train when crossing a road or railway track.

Further illusions associated with health hazards, both identified in the late 19th century, are the Ebbinghaus and Delboeuf illusions. The former was created by German psychologist Hermann Ebbinghaus and the latter by Belgian experimental psychologist, Joseph Remi Leopold Delboeuf.

Both illusions arise from our perception of relative sizes. When we look at the Ebbinghaus illusions we see two identical circles, one surrounded by larger circles and the other by smaller small circles, as shown below. Here the inner circle on the left appears smaller than the one on the right.[32]

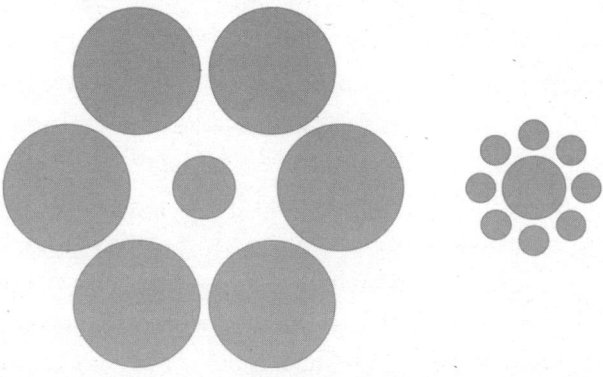

Figure 39

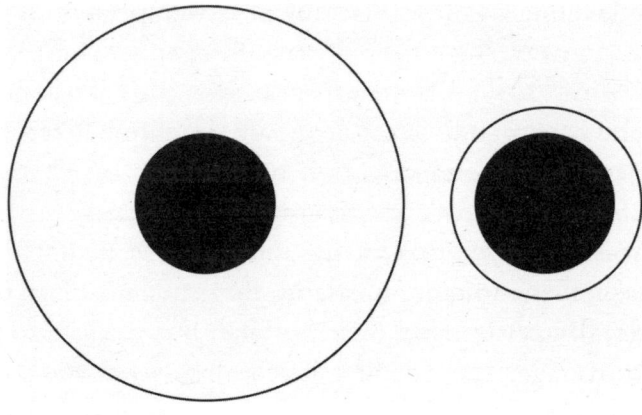

Figure 40

The Delboeuf illusion, shown in Figure 40, features a central disc surrounded by a ring that creates a similar size effect. Here the left disc appears to be smaller than the one on the right despite their both being identical.[33]

Both are illusions that influence the amount of food we eat, how much we drink, how effectively our teeth are filled, and how well certain types of surgery are performed. One study, for example, found that serving food from smaller bowls on to smaller plates significantly reduced the number of calories eaten without any diner being aware. When served on standard plates from a large serving bowl, the food eaten increased by 57 per cent.

Similarly, drinking from a tall, narrow glass rather than a short, wide one will minimise the number of liquid calories consumed. When asked to pour an exact measure without a measuring device, even professional bar staff have been found to pour up to 30 per cent more into wide glasses than thin ones. In our study, those given a wide tumbler poured twice as much as those with a narrow glass (397ml/14fl oz compared to 796ml/28fl oz). Similarly, giant bottles of cola have been shown to increase the amount people serve and consume by up to 45 per cent.[34]

No less important is the role of this illusion when you go to the dentist. In a test of the illusion's effect, Robert O'Shea and his colleagues gave eight, practising specialist dentists twenty-one isolated teeth each, randomly sampled from a collection of extracted teeth. The teeth contained holes and the dentists were instructed to prepare a cavity in the root end of each tooth to receive a filling. The researchers emphasised they should cut conservative (minimal) cavities.

On average, they drilled holes almost *twice* as large as necessary in half of all the teeth. The rounder the tooth, the greater their error was likely to be.

As O'Shea points out: 'If it is accepted that visual illusions can influence treatment in dentistry, then they may influence treatment and diagnosis in other health care ... We chose a model system resembling the classic Delboeuf illusion with rigid, near-circular shapes ... Similar displays are common in medicine and surgery', and 'for a health practitioner, underestimating the size of something, be it a vessel, a tumour, or other lesion, could have serious consequences.'[35]

Similar problems have been encountered in many branches of medicine. In laparoscopy or keyhole surgery, a miniature camera and small incisions are used to examine or treat conditions in the abdomen or pelvis. The camera lets surgeons see images on a computer screen as they diagnose and treat conditions, remove organs or take samples for biopsy.

Keyhole surgery offers several advantages over traditional surgery by shortening recovery time, causing less pain and reducing the risk of complications, while involving risks based on the Delboeuf illusion.

In one type of laparoscopy operation, bile duct surgery, 97 per cent of errors were attributed to such 'perceptual errors' and only 3 per cent to a lack of surgical skill. So overwhelming was their misperception that most surgeons refused to believe they had a problem.[36]

Visual Blind Spots

Vision specialist Richard Gregory termed the Kanizsa triangle seen earlier (Figure 28) a 'cognitive fiction' produced by how the brain fills gaps in everything we see. Indeed, we experience 'cognitive fiction' constantly throughout our waking hours. At the point where optic nerves join the eyeballs, an absence of photoreceptors means we are blind in that part of our eyes. Yet we never see blank spaces and we become aware of them only under exceptional circumstances.[37]

Put this to the test by closing your right eye and holding this book about 50cm away. Keep looking at the playing cards in Figure 41 as you slowly bring the page closer.

Figure 41

At a certain distance, the cards will vanish as the image falls on to your blind spot, and your brain fills the blank space with white to match the surroundings, a process known as 'perceptual completion'. You can do the same by closing your left eye and bringing the book closer, which will cause the rabbit in the hat to disappear.

Quirks of our visual system also give rise to Mach band illusions. The ability to spot predators lurking in shoulder-high grass could have meant the difference between living and dying for our hunter-gatherer ancestors. Survival would often have depended on rapidly and efficiently detecting where the grass ended and the outline of a predator began.

In 1865, Austrian physicist Ernst Mach noticed something strange at the point where darker and lighter tones meet.

Notice the darker stripe in the region marked A and the brighter one at B where darker and lighter tones meet. Although they appear real enough, both bands are, in fact, illusions. Our eyes constantly move from one point in our

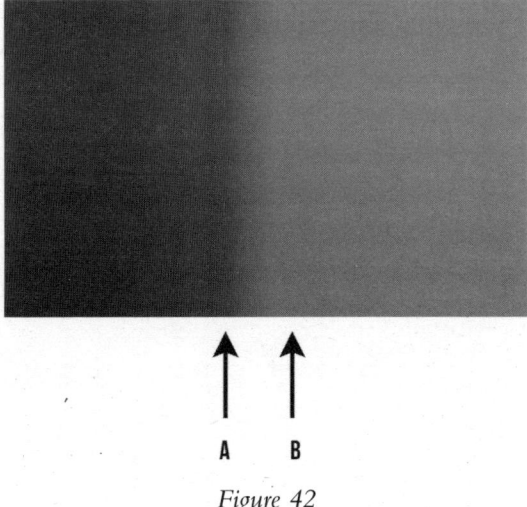

Figure 42

surroundings to the next, fixating on any point for as little as 200ms. Because of rapid changes in light levels and contrasts, the brain perceives these bands even where none is present.[38]

While Mach bands may seem inconsequential, some 40 million errors a year are made when reading X-rays, and many of these mistakes are attributed to this illusion.[39] Mach bands are often found in chest radiographs, most frequently along the spine. Those overlapping with bone can be mistaken for fractures, while any caused by skin folds can mimic the appearance of air in the space surrounding the lungs. On dental X-rays, they can give rise to a mistaken diagnosis of cavities.[40]

Pareidolia

In 1977, Angelica Rubio's mother was cooking tortillas in the small Mexican town of Lake Arthur when 'the face of Jesus' appeared on one of them. A local priest proclaimed it a 'sign from God', word of the 'miracle' quickly spread, and her modest home became an international tourist attraction as the devout flocked to it in their thousands. There, they knelt to pay homage to the burnt flour tortilla. Carefully preserved in a glass-fronted box and surrounded by flowers, it enjoyed pride of place in the front room.[41]

Humans are particularly good at projecting familiar sights on to nebulous objects such as clouds, shadows, ink blots, faces, body parts, animals, an overbaked tortilla, a slice of slightly burnt toast or some other real-world object. This illusionary blind spot, pareidolia, has intrigued scientists and writers for centuries.

In *Hamlet*, the Prince of Denmark and his courtier, Polonius, have the following exchange.

> *Hamlet:* Do you see yonder cloud that's almost in shape of a camel?
> *Polonius:* By th' mass and 'tis, like a camel indeed.
> *Hamlet:* Methinks it is like a weasel.
> *Polonius:* It is backed like a weasel.
> *Hamlet:* Or like a whale.
> *Polonius:* Very like a whale. (*Hamlet*, Act 3, Scene 2)

Rather than seeing what is there, we interpret random shapes as something more familiar, most frequently as a face. 'We've over-learned human faces,' says cognitive neuroscientist Takeo Watanabe. 'The downside is if you learn something too well you may not see what is really there.'[42]

This is often because the fusiform gyrus, a brain region responsible for recognising faces, has two sides. The left side performs a fast, intuitive judgement of whether something is a face, while the right side invests more time and effort to arrive at a slower but more accurate assessment. To save energy, your brain often jumps to rapid conclusions, if sometimes incorrectly.

'Finding meaning in ambiguous stimuli depends on conceptual evaluation and cortical processing events similar to those typically observed for known objects,' explains neuroscientist Joel Voss. 'To the brain, the vaguely Elvis-like potato chip truly can provide a substitute for the King himself.'[43]

This illusion differs from the others, however, in that its presence in medicine sometimes proves an aid rather than a handicap to diagnosis. 'Pareidolia is relatively common in radiology,' says consultant radiologist Mark Radon, 'perhaps because it aids memorisation and pattern recognition, and maybe because it is entertaining.'[44]

Frank Gaillard, founder of Radiopaedia.org, lists dozens of such pareidolias, including those associated with fish, birds, animals and insects. They include the scallop sign of rheumatoid arthritis, the manta ray sign indicative of a bladder problem, the giraffe pattern suggestive of a thyroid problem, and the elephant-on-a-flagpole sign indicating an enlarged renal pelvis. One often seen in textbooks is the 'butterfly' glioma, a fatal form of brain cancer (see Figure 43).[45]

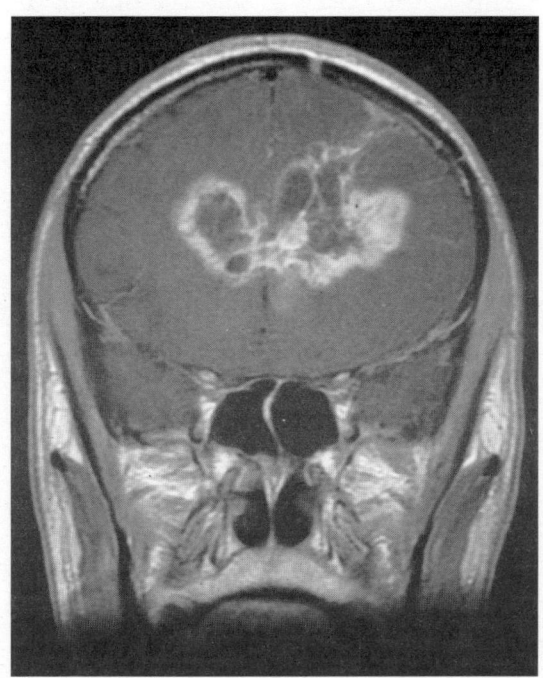

Figure 43: The 'Butterfly Glioma'.

While sometimes helpful, especially to trainees, pareidolia illusions can also mimic lesions. This can cause radiologists to report non-existent pathology, resulting in unnecessary workups or more invasive procedures. Thus, knowing how to detect and handle these illusions may help prevent premature or incorrect diagnoses. 'Perceptual errors in radiology ... are a significant contributor to patient harm,' say health scientist Robert Alexander and his colleagues. 'Yet ... the techniques taught, while valid, do not result from an ... understanding of the human eye–brain system.'[46]

While our ability to make sense of apparently random data often leads us astray, there are occasions when it helps us transform a seemingly nonsensical pattern into one that makes perfect sense. Take, for example, the confusion of shapes shown in Figures 44 and 45.

These images were part of our research into how people recognise 'emerging images'. In this study, we tracked people's eye movements as they tried to find the hidden objects. Our results showed that even though it took between three and ten seconds for people to identify what they saw, their eyes went to the right spot within half a second. This suggests that our brain can quickly detect the object's location, even when unaware of what we are looking at.[47]

Figure 44

YOU COULD HAVE FOOLED ME!

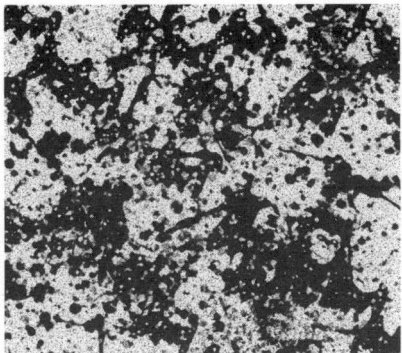

Figure 45

To test the role of conscious awareness, people were given hints about what was concealed. While this enabled people to recognise hidden objects faster, their eye movements remained the same. This tells us that the clues mainly sped up the final decision stage, not the underlying visual search.

Figure 46

When bioengineers Gert van Tonder and Yoshimichi Ejima showed people an image similar to Figure 46 for the first time, most reported a blob-like bulge in the middle of the image, a roughly spherical body. This was interpreted as a bird, a lion cub, a dog with a tiny head, a funny-looking bear, a cow with a big head, a jogger stretching out, an iguana, and two strange elephants.[48] Only one in twenty people got it right on a test we conducted. (Answer at the end of this chapter.)

Studies like these reveal the back-and-forth connection between fast, unconscious perception and slower conscious recognition. They show that our visual system detects objects exceptionally quickly. But turning that signal into a conscious concept takes longer. These findings help explain the subtle interplay between seeing and recognising objects in the world around us and reveal much about how your brain seeks to make sense of such a jumbled assortment of dots, blobs and patches.

'Pareidolia doesn't distinguish between something or nothing,' explains computer scientist Laura Lanford, because the strategy offers an evolutionary advantage. 'If a person sees a bear and a bear is indeed there, the person can run away. If the person didn't see a bear and there really was none, it would be no big deal. But if the person didn't see a bear and there was one, this would be a big problem. As such, it's better to mistakenly think there is a bear when there isn't, and not get eaten'.[49]

Cross-Modal Illusions

Since the days of Aristotle, schoolchildren have been taught that we possess five senses – sight, hearing, smell, touch, and taste. These were believed to function independently and could be safely studied in isolation.[50]

Both are untrue.

As well as the five customarily recognised senses, there are at least 16 others, including *thermoception*, an ability to sense heat and cold; *proprioception*, the sense that informs us about where parts of our body are located in respect to other body parts; *nociception*, which communicates the sensation of pain; *stretch reception*, the sense of mechanical force in our internal organs; *chemoreception*, allowing detection of hormones and drugs in the bloodstream, and *magnetoception*, which detects magnetic fields. All these are in addition to the senses that alert us when we feel hungry or thirsty.[51]

And all are interconnected. Each of our senses influences the others through a process known as cross-modal interference, meaning what we see, for instance, is profoundly influenced by what we simultaneously hear, smell, touch and taste together with our experiences of pain and temperature, and so on.

You can put this to the test with the following simple experiment. Take three empty matchboxes and fill one with coins. Weighing it in your hand you will, not surprisingly, find it far heavier than the two empty ones. Now lift all three together and, suddenly and inexplicably, they all feel significantly lighter.

Why?

A likely explanation is that when the boxes are lifted together their increased size causes the brain to expect them to weigh far more and the absence of this expectation causes them to seem to weigh far less. Close your eyes and the illusion vanishes.

One of the rarest and most remarkable examples of cross-modal illusion is synesthesia, derived from the phrase 'joining of the senses'. Experienced by around 4 per cent of the population worldwide, many experts now regard it as neither an illness nor a disease but a unique ability of the higher brain regions used for language and attention.[52]

A synesthete may see graphemes — letters, numbers or symbols — in different colours, 'hear' them as sounds or 'taste' them as various flavours.[53] Professor John Harrison describes synaesthesia as a 'confusion of the senses, whereby stimulation of one sense triggers stimulation in a completely different sensory modality ... in the absence of any direct stimulation to this second modality.'[54]

Even without synesthesia, the stimulation of one sense can alter experiences associated with another. This is particularly true for cross-modal visual and auditory illusions. An everyday example is a ventriloquist chatting with her dummy, whom audiences see talking back, with the dummy's voice being heard from the dummy's mouth. Those who believe this requires ventriloquists to 'throw' their voices are surprised to learn that no such ability exists or is necessary.

When two beeps accompany a single flash, most people will see two flashes. When the same visual display is accompanied by multiple beeps, they see numerous flashes.[55] You can see this illusion in a video we created as part of our research. Go to lewisandleyser.com/how-many-flashes or scan the QR code.

One of the most extraordinary and unexpected examples of how what we see affects what we hear is the McGurk effect, discovered in the late 1970s by psychologist Harry McGurk and his graduate student John MacDonald.[56]

This cross-modal illusion occurs when watching a video of lip movements while listening to different sounds spoken by the same person. If, for example, you see lip movements indicating the speaker is saying 'da-da' while what you hear is 'ba-ba', you will hear the former ('da-da'). So powerful is this illusion that, even when aware of what is happening, you have no power to override it. You can see this illusion by visiting lewisandleyser.com/mcgurk or scanning the QR code.

McGurk and MacDonald hypothesise that the illusion arises as your brain makes its 'best guess' at understanding contradictory sensory information. This is especially true when the audio quality is poor.[57]

Insights from Illusions

The visual illusions we've explored in this chapter, from the Ponzo effect to cross-modal phenomena like the McGurk effect, reveal a profound truth about human perception: our brains, despite their remarkable capabilities, are susceptible to misinterpretation and oversight. They allow what we infer to override what we perceive.

Yet illusions are far more than just visual tricks or entertainment. They are also windows into the fascinating and complex workings of our perceptual systems, revealing how our brains actively construct our reality rather than passively recording it.

The study of illusions also has profound implications across various fields. In medicine, understanding perceptual illusions like Mach bands can prevent misdiagnosis in radiology. In aviation and driving, awareness of how our visual system can be fooled in certain conditions can save lives. Even in our daily lives, being aware of how illusions influence our eating habits, social interactions and decision-making processes can allow us to make more informed decisions.

Additionally, illusions highlight the subjective nature of our perceptions. What we see is not always what is there, and this realisation should humble us. It reminds us to approach our perceptions and judgements with scepticism and an openness to alternative perspectives.

As we move forward, let's remember that our perception of reality is, in many ways, an illusion constructed by our brains. This doesn't mean we can't trust our senses, but it does mean we should be aware of their limitations. By understanding these limitations, we can develop strategies to see beyond our perceptual blind spots, make more informed decisions and appreciate the incredible complexity of our perceptual systems. In a world where we're constantly bombarded with visual information, the ability to critically evaluate what we see becomes increasingly essential.

Illusions serve as a powerful metaphor for broader psychological blind spots that often influence our judgement and decision-making in everyday life. The unconscious inferences our brains make to rapidly process visual data, like seeing faces in random patterns, mirror the quick, often biased judgements we make in complex social and professional situations. As we saw with radiologists interpreting X-rays, even experts can fall prey to perceptual tricks, reminding us that no one is immune from visual blindness. By studying these visual illusions, we gain insight into the mechanics of vision and a valuable framework for recognising and potentially overcoming our cognitive biases.

As we learn to question what we see, we also ask what we think we know, fostering a more critical, open-minded approach to understanding the world around us. These visual puzzles offer us a unique opportunity to see beyond our perceptual limitations, encouraging us to seek diverse perspectives and make more informed, thoughtful decisions in all our lives.

In this chapter, we examined one type of cross-modality, which occurs when what we see and hear conflicts. In the next chapter, we will explore how signals from our ears, digestion and bladder can make us see the world in a very different and sometimes hazardous way.

At the end of the last chapter, we asked whether the woman sitting in a giant saucer was genuine or an illusion. It was, of course, an anamorphic *trompe d'oeil* illusion, as a side-angle view makes clear.

Figure 47

YOU COULD HAVE FOOLED ME! 121

Answers to 'Emerging Images'

Figure 48: A zebra on a road crossing.

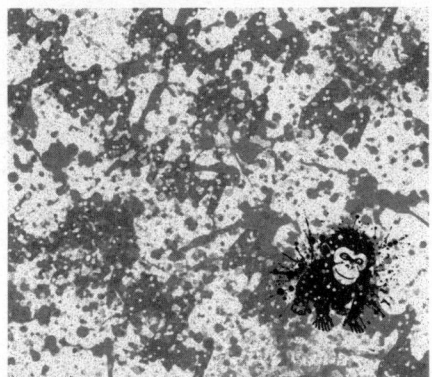

Figure 49: A chimpanzee in the jungle.

Figure 50: A Dalmatian dog.

The Perspective Illusion Study

Figure 51

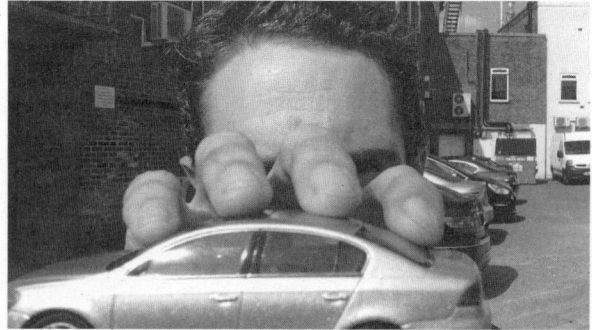

Figure 52

Figure 53

CHAPTER SIX
Blind Spots and Your Alien Brain

Your brain constantly communicates with your belly, and your belly – or more precisely your gastrointestinal tract – with your brain. Yet, aside from hunger cues, the only times you'll ever be aware of this ongoing two-way chatter are when having 'butterflies in the stomach' if you are anxious or excited, when experiencing 'gut wrenching' feelings of disgust, or when having a 'sinking feeling' in the pit of your stomach when feeling dread.

Bidirectional communications, of which we are generally unaware, occur not only between the brain and gut but with balance mechanisms in the ears, tension sensors in the bladder and even muscles in the arms. All are capable of influencing what we see and what remains unseen. According to computer scientist Douglas Hofstadter, we might explode in fear and shock if 'privy to the unimaginable frantic goings-on inside our bodies. We live in a state of blessed ignorance.'[1]

To understand the science behind gut feelings, we must travel back to 1822 and the small Michigan town of Michillimackinac.* On 6 June that year, Alexis St Martin, an 18-year-old French-Canadian trapper employed by the American Fur Company, was shot in the stomach by an accidentally discharged musket.

'The charge, consisting of powder and duck-shot, was received in the left side of the youth, he being at a distance of not more than one yard from the muzzle of the gun,' reported US Army surgeon William Beaumont, who first treated and later experimented on the youth. 'A portion of the stomach, lacerated through all its coats, and pouring out the food he

---

*The name of present-day Mackinac Island and Mackinac County.

had taken for his breakfast, through an orifice large enough to admit the forefinger.'[2]

Despite the damage, and long before the discovery of antibiotics, Alexis somehow recovered from his horrendous injury and went on to enjoy a long and healthy life, fathering several children. He was, however, left with one dreadful souvenir of the accident. A 6cm circumference hole, a permanent fistula, in his abdomen from which everything he ate and drank constantly exuded. Beaumont used this 'window' into St Martin's gut to extract gastric juices and observe the effects of emotions on his digestion.

Beaumont's research, published in his 1838 book *Experiments and Observations on the Gastric Juice and Physiology of Digestion*, made him famous in medical circles as the 'father of gastric physiology'. He was the first to identify the gut–brain axis, a two-way communication pathway between what goes on above the collar and what goes on below the lungs, proving the suggestion some thirty years later of British doctors William Bayliss and Ernest Starling who claimed the gut could work independently of the brain.[3]

Further proof came in 1917 when Ulrich Trendelenburg, a German doctor, mounted a loop of guinea pig intestine on a hollow tube inside a bath of warm, nutritious solution and supplied it with oxygen. When he blew into the bowel, the gut blew back. The conclusion drawn from this simple experiment was profound. The guinea pig intestine must have detected Ulrich blowing into it to produce a coordinated wave of contraction and relaxation despite being severed from a brain.[4] The gut has a 'mind' of its own.

'Remember Pinocchio?' asks Professor John F. Cryan. 'It's the story of who is in control. Often, the person who is really in control is the partner of the person who thinks they are in control. We've been so focused on the brain, but perhaps it is the microbiome pulling the strings.'[5]

Only in the past twenty years has the extent to which blind spots are controlled by the trillions of microbes inhabiting our intestines – what may fairly be described as our 'alien brain' – been recognised. 'The gut microbiome is home to the most extensive collection of microorganisms in the human body,' says researcher Thomaz Bastiaanssen. 'It encompasses the trillions of bacteria, viruses, fungi, and other microorganisms that live inside the … stomach but also the mouth, oesophagus, pancreas, liver, gallbladder, small intestine, and colon. After the brain, the gut contains the body's largest number of neurons.[6]

It's been estimated that we have between 30 and 400 trillion microorganisms in the gut and from three to 100 times more bacteria in the gut than there are cells in the human body.[7]

Each of us possesses a unique collection of around 160 different types of bacteria, which means you probably have different bacteria compared to the person sitting next to you.[8]

This microbiome performs various functions, from manufacturing vitamins and regulating the immune system to protecting the gut's integrity. However, a number of impacts on our brain due to the brain–gut axis have also been discovered.

Some bacteria in our gut, for instance, make oxytocin, which encourages increased social behaviour, along with substances that make us calmer under stress. They have also been shown to influence vulnerability to diseases, including Alzheimer's and Parkinson's, and developmental conditions such as autism. A substance called synuclein, present in the brains of those with these conditions, is made by gut bacteria and travels upwards to the brain.[9]

The microbiome also metabolises the amino acid tryptophan into serotonin,[10] which plays a vital role in both the brain and gut, where low levels cause depression and constipation respectively, conditions leading to around 3 million doctor consultations and 100,000 hospital visits each year.[11]

Conversely, what is happening in our brains can also have a major impact on our gut. Fear-inducing and stressful situations lead to cortisol release from the adrenal glands, which slows the muscles controlling gastrointestinal peristalsis and the production of digestive chemicals in the stomach and small intestine.[12]

The 'brains' in our head and our belly are in constant two-way communication. As a result, anything that affects our gut brain also influences our head brain, and vice versa.[13] The primary communication route between the two 'brains' is the tenth cranial nerve, also called the vagus nerve, from the Latin for 'wandering'. After emerging from the base of the skull, the nerve meanders through the body with branches to the heart, lungs, larynx, stomach and ears. As part of the parasympathetic branch of the autonomic nervous system (ANS), the vagus nerve carries information about the digestive system to and from the brain, regulating gut motility, local blood flow and enzyme secretion. Some 20 per cent of the vagus nerve fibres are dedicated to this two-way traffic between the brain and the microbiome.

Some parts of the gut are only able to exert positive effects on mental health if this nerve is working correctly. When communications break down, the system becomes 'deregulated', resulting in, among other conditions, abdominal pains, which can cause distraction and can lead to inattentional blind spots.

Up to the moment of birth, we are all 100 per cent human. From then on, we become 10 per cent human and 90 cent microorganism, with our microbiome containing around 150 times more genes than our human genome.[14] While multiple factors influence a baby's microbiome during the first year of life, the way the infant was delivered is one of the most significant. Infants born by caesarean section have significantly different early gut microbial features compared to vaginally delivered infants, including lower diversity and richness, and the prevalence of certain bacteria species.[15]

Other factors influencing a child's gut biome include diet, hygiene, social behaviours, genetics, other children, parents and pets in the home environment, whether or not they receive antibiotics and if they were breast- or bottle-fed.[16]

Breast milk is unique in helping establish a healthy microbiome because among its most important ingredients are milk oligosaccharides (HMOs). Consisting of some 200 indigestible proteins, these HMOs protect the baby's gut against harmful microbes, which mistake them for sugars and bind to them rather than infecting the child. HMOs also nourish beneficial bacteria, encouraging their colonisation of the intestine. Despite over a century of trying, the manufacturers of formula baby foods have been unable to create artificial breast milk that replicates these microbial benefits.[17] Within a baby's first year, between 500 and 1,000 species of bacteria will form, with some passed down from one generation to another like family heirlooms.

As children grow into adults, their lifestyle choices, especially concerning sexual behaviour, dramatically influence the type of bacteria in their microbiome. *Segatella* bacteria, for instance, which help break down dietary fibre and are associated with greater cardiovascular health, are found in the guts of 70 per cent of gay men who practise condom-free anal sex with multiple partners. In comparison these bacteria are only found in 20 to 30 per cent of the general population in Europe, though in 90 per cent of non-industrialised populations, such as those living in the Amazon or parts of Africa, most likely due to their very different diets.[18]

Why there's such a difference between gay men and the general population when it comes to levels of *Segatella* is still unclear, but it illustrates the intimate connection between lifestyle and our second brain and how both influence what we see and how we behave. This also applies to another way of satisfying our appetite: eating.

Hungry Judges and Starving Juries

For centuries, judges who were impatient to obtain a guilty verdict used hunger to secure a conviction from reluctant jurors. In 1293, a judge was angered when a jury produced a verdict he disliked and informed them that they would 'stay shut up without food or drink till tomorrow morning'.[19] This led to the 17th-century poet Alexander Pope's sardonic comment: 'The hungry Judges soon the Sentence sign, And Wretches hang that Jurymen may Dine'.[20]

In Charles Dickens's *Pickwick Papers* (1837), Pickwick's companion, the self-styled 'romantic poet' Augustus Snodgrass, says to his lawyer Mr Perker: 'I wonder what the foreman of the jury, whoever he'll be, has got for breakfast?' 'I hope he's got a good one,' Perker replies. 'A good, contented, well-breakfasted juryman is a capital thing to get hold of. Discontented or hungry jurymen, my dear Sir, always find for the plaintiff.'

Recently, the debate has shifted from the likely impact on verdicts reached by hungry jurors to decisions made by hungry judges. In theory, the law should be applied without fear or favour, with the fate of the accused determined solely by the facts of a case. Lawyers, however, have long suspected judges are often influenced not so much by objective reasoning as the length of time since they last ate.[21]

A team of researchers led by Shai Danziger suggest there is some merit to that suspicion.[22] After analysing over a thousand judicial rulings on parole appeals, the researchers found judges granted 65 per cent of parole requests at the beginning of the day's session and almost none at the end. Immediately after lunch, approvals again returned to 65 per cent. Danziger and his colleagues concluded meal breaks might restore mental resources, returning the chance of their granting parole to the initial level. 'Our findings support the view that ... legally irrelevant situational determinants – in this case, merely taking a food break – may lead a judge to rule differently in cases with similar legal characteristics.'[23]

It has also been demonstrated that the microbiota not only protects the gut against viral diseases but influences the release of molecules that improve mental performance.[24] This suggests the effectiveness of our cognitions and actions could be enhanced using probiotics to improve gut microbiota. In one study, enhancing the microbiome in this way increased the ability of participants to process, apply, remember and ultimately learn from new information.[25] Gut bacteria health has also been shown to influence how emotional, bold or anxious we are, our perception of pain, how we respond to stress, and the likelihood of blind spots causing us to look without seeing.[26]

But communication between gut and brain is not just one-way. The brain also exerts a powerful influence on gut bacteria. Even mild stress can tip the microbial balance in the gut,[27] triggering a cascade of molecular reactions that give feedback to the central nervous system.[28]

While the importance of the brain–gut axis has been recognised for decades, researchers have only recently realised that another part of the body, the balance mechanisms in our inner ears, is also an active and highly influential contributor to this two-way communication network. The messages these send to and receive from the brain play a pivotal role in whether or not blind spots occur.[29]

Balance-Driven Blind Spots

As FlyDubai Flight 981 approached Runway 04 at the Russian airport of Rostov-on-Don on 18 March 2016, the pilots knew they were in for a challenging landing. The pitch-black night was swept with torrential rain. Winds gusted up to 65km/h (40mph). Low cloud meant poor visibility. Wary of wind shear, a sudden tailwind that can cause an aircraft to drop out of the sky, the crew opted for a 'go-around'. Increasing engine power, they climbed away from the runway.

The Boeing 737 flew in circles for ninety minutes as the pilots waited for a break in the storm. When none came, and with fuel running low, they decided to make a second attempt at landing. Once again, the strong winds forced them to abandon the attempt and head back into the sky. Moments later, the plane came zooming back down out of the clouds at a steep angle. Striking the runway, it erupted in a fireball and all 62 passengers and crew died instantly.[30]

Understanding what went so catastrophically wrong offers insights into the creation of blind spots that have nothing to do with the five causes – inattention, change, choice, expectations and illusions – discussed so far.

After flying around the airport, burning through most of its fuel, the plane was lighter. As a result, on entering the clouds and throttling the engines to full power, the plane accelerated with unusual alacrity. This caused the two pilots to rise in their seats like they were on a rollercoaster, zooming over the crest of a steep loop. The abrupt movement affected the delicate balance mechanisms in their inner ears. Even though their aircraft was safely positioned, with limited visibility, they were convinced it was still heading skywards and about to stall. In an attempt to correct this, they pushed the stick forward to lower the nose, causing them to speed towards the ground. They emerged from low cloud in a 50-degree vertical dive, travelling at over 600km/h (370mph), in what pilots call a 'graveyard' spin as their brains misinterpreted changes to their balance mechanism as an upward, rather than a downward, acceleration.

It was a similar spatial disorientation blind spot that most likely resulted in the crash of a light aircraft piloted by John F. Kennedy Jr, which claimed the lives of all on board. On the evening of 16 July 1999, Kennedy, his wife Carolyn and her sister Lauren took off from New Jersey's Essex County Airport en route to Martha's Vineyard. The Piper Saratoga aircraft made an overly sharp turn and fell from the sky while flying over a stretch of water about 65km (40 miles) from the landing strip. Nose down, it hit the water at a speed of over 85km/h (53mph).[31]

It is probable, if never proven, that the bodily driven blind spot that killed Kennedy Jr was 'the leans'. This occurs following a gradual and prolonged turn, which goes unnoticed by a pilot because it is below the detection threshold of their semi-circular canals. On returning to level flight, affected pilots believe their aircraft is banking in the opposite direction and lean in the direction of the original turn to regain their visual perception of a vertical posture.[32]

To understand how this happens, we need to examine the inner ear mechanisms more closely, as illustrated below.

For a 3D representation of the image, go to lewisandleyser.com/ear or scan the QR code on the next page.

The ear has a hearing component, the cochlea, and a balance component, the vestibular apparatus. The vestibular apparatus works with the visual system and information from

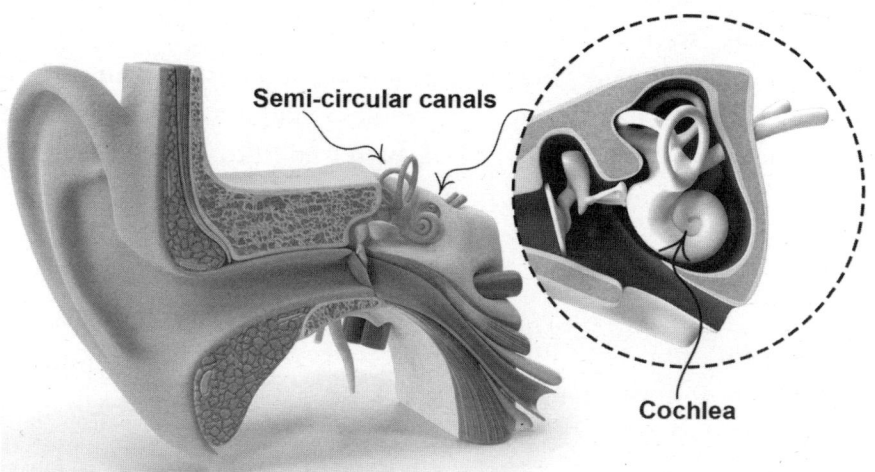

Figure 54

receptors in our joints and muscles to enable us to maintain balance when moving our heads. It comprises three semi-circular canals at right angles to each other. One is horizontal and one vertical, while the third travels from the front to the back of the body. Together, these three axes can detect motion in any direction via a gelatinous fluid called endolymph within each canal. This fluid moves (relative to the canal) when our body moves. In doing so, it bends fine hairs – cilia – lining the canals and sends signals to the vestibulocochlear nerve.

The endolymph remains stationary while the appropriate semi-circular canal rotates each time we nod or turn our heads. However, if your head continues to move – as in an aircraft, on a ship or while riding a roller-coaster – the endolymph too moves, creating a false sensation of no longer turning. In a situation such as landing at night, the lack of visual cues especially below the aircraft can make the runway appear tilted and sloping. As a result, a pilot may descend too rapidly, follow an incorrect glide path, and risk a crash landing. 'When you initiate the go-around and still have some visual reference, you're fine,' explains aviation analyst Jerry Wise. 'But once you get into the clouds, your senses start to play on you.'[33] Unless recognised immediately, such blind spots can lead to a mid-air collision, an aircraft stall or what aviation experts somewhat paradoxically term a CFIT – controlled flight into terrain.[34]

Such inner ear caused illusions can be costly both financially and in terms of lost lives, with more than 20 aircraft crashes occurring at a cost of $400 million in the USA each year.[35]

Motion impacts more than just our perception of orientation, however. Parents who rock their baby to sleep have long known that motion increases drowsiness. But it was not until 1976 that cardiologist Aston Graybiel and researcher James Knepton from the U. S. Naval School of Aviation Medicine identified the cause, naming it the 'sopite syndrome', from the Latin *sopire*, 'to put to sleep'.

Studying the effects of sopite syndrome on naval cadets, Graybiel and Knepton characterised the syndrome primarily by evidence of drowsiness, a decreased ability to concentrate and performance errors. The problem, they concluded, arose from a conflict between signals reaching the brain from the eyes and inner ears.[36]

In commercial air flight, because flight attendants move around far more than flight crew, have fewer external visual references to the aircraft's proper motion and lack the continuous sensory feedback a pilot or co-pilot receives while flying the plane, they experience more sopite syndrome problems, including post-flight fatigue and insomnia, than pilots.[37]

Bladder-Driven Blind Spots

Although there may not be any link between blind spots, brains and bladders, research suggests an urgent need to urinate can significantly increase *or* decrease the likelihood of looking without seeing, depending on the situation.

The volume of urine passed in a healthy person is mainly determined by diet and fluid intake. In temperate conditions, it lies within the range of 800–2,500ml per 24 hours. Bladder pressure remains constant at between 100ml and 400ml of fluid. Beyond this point, however, increased urine production causes the pressure to rise rapidly.

The first urge to urinate is felt at a bladder volume of around 150ml, with a marked sense of fullness occurring at

400ml. At the former level the urge occurs periodically and lasts from a few seconds to more than a minute. As the bladder becomes ever fuller, urination reflexes become increasingly frequent and powerful.[38]

Children's needs are even more significant since, unlike adults, who can inhibit emptying their bladder, urination in infants is a simple reflex action that occurs whenever the bladder becomes distended. Anyone who has taken a small child on a long journey by car knows how this can produce regular pleas for a 'rest stop'. High-salt, high-protein food exacerbates the situation, with many motoring snacks, such as burgers – high in both – increasing one's thirst and creating a vicious cycle of drinking, which triggers a desire to pee.[39]

All of these constitute a significant distraction for drivers, especially for female drivers, who are more likely to be accompanied by children. This assumption was confirmed by Finnish traffic researcher Ida Maasalo and her colleagues in 2019. 'Our results showed,' she writes, 'that drivers with child passengers were more often assessed as inattentive compared to drivers without child passengers, which indicates that children in the vehicle represent a potential source of distraction.'[40]

While it cannot be said that the child's desire to urinate was always the cause of their proving a distraction, it seems likely that a need to do so would increase their chances of becoming so.

While the need to urinate can sometimes be a distraction, it can also inhibit looking without seeing when making decisions. By enhancing self-control, a full bladder can reduce impulsive, short-term thinking and increase longer-term, more reasoned responses.

In a study designed to demonstrate this, consumer psychology researcher Mirjam Tuk and her colleagues asked participants to compare the taste of different types of water. One group was told to take only a few sips, while the other was asked to drink as many cups of water as possible. They were then told to choose between receiving a modest cash reward the following day or a much more generous payment in a month. To their surprise, the psychologists found the second group, those whose bladders were the fullest, were more likely to choose the larger amount later than a small amount almost immediately. This suggests that should you ever have to make a tough decision, you would be better off doing so before going to the lavatory. Your desire to pass water will likely blind you to the chances of making a poor choice while increasing the likelihood of making a good one.[41]

Arm-Bend Blind Spots

Even the most seemingly inconsequential bodily movements can influence what we see or fail to see.[42] One study compared the differences in purchasing choices between shoppers carrying hand-held baskets (requiring arm bending) and those pushing carts (requiring arm extension) and found that those carrying a basket tended to favour vice over virtue, paying more attention to immediately rewarding treats like sweets and chocolate over healthier items such as fruits or vegetables.

While it might be argued that this is because many indulgent items, like chocolates or sweets, are lighter and more easily carried than fruit or vegetables, the researchers believe arm bending activates the brain's 'reward centre', making people focus on instant gratification. This effect was more noticeable when the dominant arm, most often used to reach out for desirables, was bent. So, over time, arm bending gets linked to rewards. Such studies demonstrate that subtle body positions can sway our higher-level choices without us realising it through deep-rooted biological mechanisms attuned to pleasure. Other studies have shown that arm flexion and extension also influence cognitive processing.[43] For example, people have been found to solve problems requiring creative insight more easily when flexing rather than extending their arm.[44] Researchers have also found that arm flexing increases our attention, reduces errors and improves long- and short-term memory.[45]

Researchers believe these differences arise because bending the arms is an 'approach' response (indicating 'come to me'), while flexing, pushing something away, is associated with avoidance. 'Approach actions may activate the semantic concept "good", whereas performance of avoidance actions may activate the concept "bad",' suggests psychologist Jens Förster. 'Activation of these mental representations may then suffice to broaden or narrow the scope of perceptual and conceptual attention, irrespective of whether approach or avoidance behaviours are enacted outright.'[46]

The Power of Bodily-Driven Blind Spots

Visceral drives enormously impact our everyday lives. Blinding themselves to health dangers, people buy more food high in sugar, salt and fat when their stomachs are empty than when they are satiated.[47] People experiencing sexual arousal close their eyes to the risks of unsafe sex.[48]

'Moreover,' says Mirjam Tuk, 'in hindsight, people have a hard time fully realising the impact of these visceral states on their behaviours. The impact of visceral drives generalises beyond the visceral domain, resulting in a general increase in impulsive behaviour.'[49]

Bodily-driven blind spots typically operate outside conscious awareness. An increased heart rate, too slight to be noticeable, can make people either less or more open to visual or auditory persuasion. An empty stomach reduces our decision-making powers, while a full bladder increases them. While its importance has been recognised for decades, researchers have only recently realised the extent to which feeling satiated or hungry contributes to messages sent to and received by the brain.

What the 19th-century French physiologist Claude Bernard called our *milieu intérieur* – interior environment – affects how we pay attention, and in doing so increases our chances of falling prey to one or more of the types of blind spot described in the previous chapters.

The concept of bodily-driven blind spots reveals a new dimension to our understanding of cognitive biases. Our perceptions and judgements are not solely the product of our conscious mind but are deeply influenced by the complex interplay of our bodily systems. As we move forward, we must remember that our mind and body are not separate entities but parts of an integrated system. By recognising the bodily sources of our cognitive blind spots, we can develop strategies to mitigate their effects. This involves being more mindful of our physical states when making important decisions or considering how our lifestyle choices influence our thought patterns through their impact on our microbiome.

Ultimately, understanding the 'alien brain' within us provides fascinating insights into human biology and offers practical tools for improving our decision-making, enhancing our well-being and gaining a more complete understanding of ourselves. As we unravel the mysteries of the mind–body connection, we may find that the key to clearer thinking lies not just in our heads but throughout our entire body.

CHAPTER SEVEN

Do You See What I See?

Put down this book for a moment and look around you. Maybe you are in a room, a library, a garden, the countryside or on the beach. Perhaps you are travelling by train or by air. Wherever you are, study your surroundings. Note the shapes, textures, colours and arrangements of things. Identify the various objects and their relationship to one another, the interplay of light and shadow.

Now, ask yourself this: Is what I am seeing reality,* or is it what philosopher Alva Noë has termed a 'grand illusion'.[1]

An increasing number of neuroscientists and philosophers believe it to be the latter.

As visual attention researcher Jeremy Wolfe puts it: 'Everything that you have ever seen, felt, thought, tasted, imbibed, was interpreted by a sub-set of your brain that ... builds a grand illusion of the world outside. And that illusion, that simulation of the world, is the only thing you have ever interacted with or ever will.'[2]

The illusionary nature of our perceptions is hardly a recent idea. As early as 1752, French scientist Pierre Louis Morceau de Maurpertuis speculated that our perceptions 'are properties of the mind alone in the universe'.[3] Philosopher Nick Bostrom has taken this idea even further by suggesting that everything we see, hear, taste and touch may arise from a computer simulation created by the alien equivalent of a bored teenage hacker![4] Indeed, what we take to be reality is created and instantiated entirely by brain processes. Every

---

* We are using the term here as reflecting everyday perception of our surroundings, while being aware that, in theoretical physicist Carlo Rovelli's words, 'reality is not what it seems'.

sensation is generated by a very special kind of neural circuitry.

Such an idea turns on its head everything most people implicitly believe true. It places us in much the same position classical physicists found themselves in during the 1920s and 30s at the dawn of quantum physics. In his book *The Age of Uncertainty*, German author Tobias Hürter writes: 'Imagine one day, you found out that the world you live in works completely differently from the way you thought. The buildings, streets, trees, and clouds are nothing but pieces of the theatre scenery shifted by forces you never dreamed existed.'[5] While that may sound like a scene from *The Matrix*, it's precisely what happened to physicists a century ago. They were forced to accept a more profound 'reality' lay behind the concepts and theories through which they saw the world and began arguing whether speaking of 'reality' still made any sense.

The Anatomy of Seeing

To help us better understand how the world we see may not be an objective reality captured by our eyes but a subjective construction created by our brains, let's go inside our eyes to discover how they transform photons of light into organised images.

Eyes have evolved into various shapes, sizes and positions on the skull across species. Insects have compound eyes, offering a panoramic view with a large field of vision. Deep-sea fish have evolved tubular eyes that help them detect prey and predators swimming above them. Cows have sideways-facing eyes that can see 360 degrees, which helps avoid predators, yet their depth perception is poor despite this wide range of vision. They cannot distinguish a hole in the ground from their own shadow. To them, both appear the same.[6]

Not only are there differences in the size and shape of eyes across species, but there are also differences in the

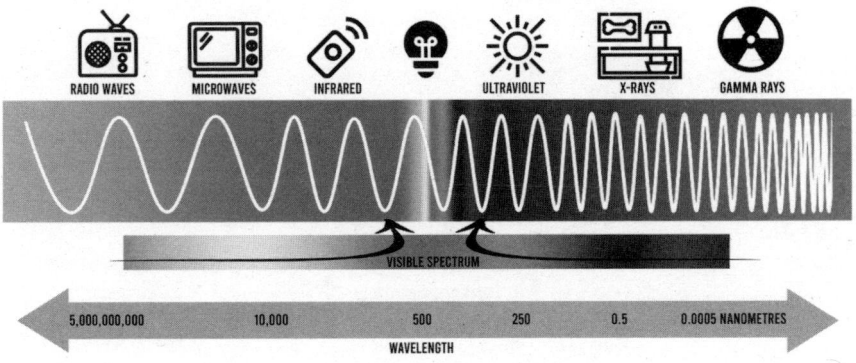

Figure 55

wavelengths they are able to detect. Figure 55 shows that human eyes only detect a narrow band of wavelengths between red at 380 nanometres (nm) and blue at 700nm. If we could see a different part of the visual spectrum, like some animals and insects, our perceptions of the world would radically change.

Pythons, rattlesnakes and goldfish see into the infrared, and frogs, bees, and mosquitoes into the ultraviolet. Swatting a housefly proves nearly impossible because they see things four times faster than we do.[7] Yet despite these differences, the eyes of every species undertake the same essential task: converting light energy into chemical changes and electrical signals for onward transmission into the brain.

As with other sensory and cortical systems, the ability to see starts taking shape within the first months after conception. Retinal cells, rods, cones, amacrine and ganglion cells start forming around day 40 of gestation and the eyes have a full complement of cells by 160 days, although they continue to mature during the first year.[8] The muscles responsible for moving the eyes also develop soon after birth, as do the brain regions controlling these muscles, the superior colliculus and brainstem.

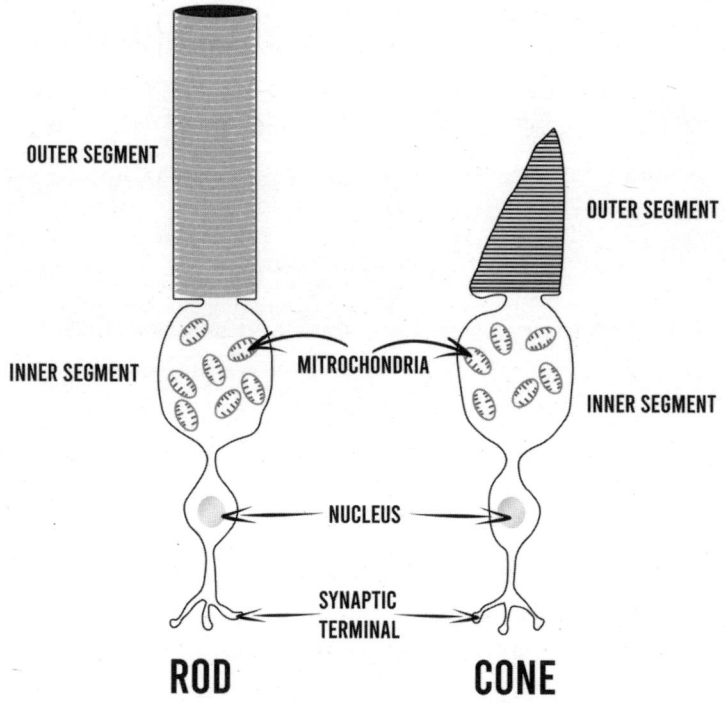

Figure 56

In sight, the lens focuses incoming light on to the fovea, which is the area of the retina with the highest concentration of photoreceptors. Seeing detailed information about different parts of the world is made possible by moving the eyes to various points in the visual scene. Since a newborn baby has less developed visual apparatus, everything he or she sees will have a slightly blurry, hazy and sluggish appearance, with less ability to resolve fine detail and differentiate shades of brightness or colour than adults.[9]

Protected by layers of thick membranes and the bones of the skull, brains inhabit a world of almost total darkness. The only visual information they receive about the world comes via two types of light receptors in the eyes named after their appearance: rods and cones.

There are around 120 million rods in each human eye. Located mainly around the retina's edges, they are responsible for low-light (scotopic) vision. Unable to make out fine detail, rods see the world in black, white or grey. They are, however, incredibly sensitive to light. It has been estimated that they can detect a candle flame 30 miles (48km) away on a clear night.[10] If you want to see something on a dark night, try viewing it out of the corners of your eyes where there are the greatest number of rods.

Conversely, cones function best in bright light (photopic vision), allowing us to see in colour and perceive fine detail. There are three types of cones, each containing a slightly different pigment. Some, only responding to short wavelengths of light, enable us to see blue. Others sensitive to medium wavelengths see green, while long-wavelength-sensitive pigments see red. From these three, every other colour is created. White, for instance, indicates the presence of all three colours, and black indicates their absence.

When photons find their way through the tangle of blood vessels and nerves in front of these light receptors, they cause a chemical change in the receptors' pigments. This then triggers an electrical signal that travels through those same blood vessels and nerves to exit the skull via the optic nerve on its way to the brain.

The total number of rods exceeds the number of cones by approximately 4.5 million. The one exception is a small area in the retina's centre, measuring some 1.2mm across. This is the *fovea centralis*, Latin for 'central pit', arguably the most crucial square millimetre in the entire human body because it is the only region of the eye that sees the world in focus. Everything seen outside the fovea lacks sharpness and clarity. Occupying just 1 per cent of the retina, the fovea requires 5 per cent of the brain's visual processing capacity.

The differences in vision between the fovea and the rest of the eye have even been used in art. For centuries, people have struggled to understand the expression on one of the world's most famous faces, Leonardo da Vinci's *Mona Lisa*.

Figure 57

A chance observation by neurobiologist Margaret Livingstone revealed an intriguing difference between images falling on the cone-filled fovea and those striking the rod-lined periphery.[11] If you focus on her hands or the background, so that the face is only seen out of the corner of your eye, thus employing your low-resolution rods, the mouth appears cheerful. Yet look directly at the face using your high-resolution foveal vision, and her smile disappears!

Not only is what we see different depending on where we focus but images arriving on the retinas are upside down, blurred and constantly moving. Yet for most people the world is colourful, the right way up, sharply detailed and stationary – not because of what our eyes see but because of how our brains process incoming images.

Because the fovea is the only region of sharp focus, we have a limited useful field of vision (UFOV). Keep your head and eyes perfectly still, and your view will be limited to a horizontal field of about 210 degrees left to right, and a vertical angle of about 15 degrees. Everything outside the middle 5 degrees, horizontally and vertically, will be in low resolution. This means our eyes must constantly move around in a series of rapid jerks to capture detailed snapshots of our surroundings, which the brain integrates into a coherent understanding of the environment.[12] Known as saccades (from the Middle French word *saquer*, 'to pull' or 'to draw'), these range from the small movements made while reading to far larger ones made while gazing around a room.

The way we read, whether online or off, is seldom linear. Although most saccades progress with the text, readers do not necessarily read every word. Short words are often ignored, while longer ones may receive more than one 'fixation'. That is to say the eyes remain stationary for a brief moment, typically lasting between 150 and 300 milliseconds (ms). When reading, for example, fixation time is shorter for familiar words and longer for unfamiliar or complex terms.[13]

Rather than starting at the beginning and going through to the end, our eyes pinball around, scanning words in a tempo of saccades and fixations. Between each 'jump', the eyes take a break for between 200 and 300 milliseconds, during which they either rest or focus on a line of text to read it more carefully.[14] Approximately 10–25 per cent of saccades move the eyes in the direction opposite to word order (termed regressions). 'A covert processing mechanism thus appears to intervene between linguistic processes and overt eye movements,' report psychologists Ulrich Weger and Albrecht Inhoff. 'The critical function of this mechanism is to direct attention and linguistic processing to relevant words in the text.'[15] While we can move our eyes voluntarily, saccades occur involuntarily all the time our eyes are open, even when our view is fixed on a target. When your eyes scan this line of text, letter shapes and the space inside and around each character send snapshots to your brain, enabling it to create a mental picture of what it sees. Because this is a skill we develop in the first few years of life, the complex process becomes automatic. When we encounter the same words day after day, they can instantly be identified through eye movements.

To accumulate all this information, our visual system has to overcome the numerous challenges posed by each eye movement. First, it has to link the appearance of objects before and after a saccade. Doing this requires either information on the eye's position or specific landmarks that are the same before and after the saccade. Second, eye movements must be hidden from our awareness to prevent confusion with object movement and to avoid motion sickness. Third, because attentional resources are limited, the brain must reallocate these resources with each eye movement.[16] The fact that we remain consciously unaware of either these movements or our brief periods of blindness is a testament to the brain's abilities.

Eye tracking has been used to reveal the direction of saccades and the amount of time spent paying close attention to one aspect of the scene – the periods during which a particular area was directed to the fovea for detailed scrutiny. In a laboratory-based study, we showed young male and female subjects photographs of two pairs of very different but, we thought, equally appetite-driven things. The men saw a swimsuited female model and over-filled toasted sandwiches with French fries. The women chose between a chocolate cake and a hunky, topless man. As they looked at these images, we used an eye tracker to record the directions in which their eyes moved and the length of each fixation.

The results were surprising. While the young men (average age 21) spent more time studying the model than the food, women of the same age showed more interest in the cake than in the man.

That such gaze differences occurred will, for many, seem unsurprising. It is only natural, they may argue, that young men should notice an attractive, scantily clad girl more than a bacon sandwich and chips, a conclusion supported by academic researchers.[17]

But why were the women more interested in the cake? This difference may be because we chose a chocolate cake as their alternative image. Research has shown that women have a far greater preference than men for 'comfort' foods such as chocolate, sweets and ice cream. Research by psychotherapist Terhi Tuomisto and colleagues found that 70 of 72 self-selected 'chocolate addicts' were female,[18] while in another study, 92 per cent of these self-identified 'addicts' were women.[19]

This is not to suggest heterosexual women have less interest in a man's physique. Research by social psychologist Madeleine Fugère revealed that young women looking for boyfriends, and mothers seeking them for their daughters, considered a degree of attractiveness an essential criterion in a potential mate.[20] Provided a would-be partner was moderately attractive, however, both daughters and mothers would favour the one with the most desirable personality.

The same cannot be said of men, who are more aware of, or more willing to admit, that a woman's looks matter more to them than her personality. This emphasis may have a biological basis, as many men associate physical attractiveness with fertility.[21]

Other factors that catch and hold our attention are novelty and movement. Anything new, different or unusual, especially anything that moves, instantly catches our eye and will be intensively studied. There are good evolutionary reasons for this since such things could threaten our survival.

One hint that we see the world in discrete 'frames' rather than a continuous and unending flow came at the start of the last century with the arrival of cowboy films.[22] In these, stagecoach wheels can be seen to slow down or even turn in reverse while the coach itself continues to move forward. Known as the 'wagon wheel' illusion, this effect can also be seen in movies showing spinning aeroplane propellers, helicopter blades, jet engine fans and similarly radially patterned rotating objects. The illusion arises because, as the wheel rotates, the camera takes a sequence of snapshots. If the wheel rotates faster than the speed at which the camera takes the shots, around 24 pictures each second (the human eye perceives 10 to 12 frames per second as distinct images), it can appear as if each spoke has rotated a small distance backwards in every frame.

You can recreate the illusion by cutting a paper disc the size of a long-playing record and drawing in spokes or other radial patterns. The 'wagon wheel' reversal will occur when spun at between 33 and 78 revolutions per minute.[23] To watch it in action, go to lewisandleyser.com/wagon-wheel or scan the QR code.

The fact that we can see this illusion naked eye suggests that, like a camera, the brain also naturally slices our visual perception into a succession of snapshots. As in a movie, these still images link to make us believe we are seeing continuous motion.

In 2006, the neuroscientist Rufin van Rullen recreated the illusion in his lab. When he spun a wheel at certain speeds, all subjects reported seeing it turn the 'wrong' way, allowing him to establish that we see the world at a rate of about 30 pictures per second. The next question he wanted to answer was, what part of the brain sets this rate? To find out, he placed electrodes on his subjects' scalps, recorded their brain waves and identified a specific rhythm that rises and falls at around the correct frequency in a region associated with vision, the right inferior parietal lobe (RPL). In a later study, van Rullen used a non-invasive technique known as transcranial magnetic stimulation to disrupt the regular brain waves in the subjects' RPLs. This, he found, significantly reduced the probability of seeing the illusion. 'The subjects could still see the regular rotation of the wheels, however,' he reported, 'probably because other brain regions that don't operate at 13 Hz took over some of the motion detection.'[24]

A 'Little Man' Inside Our Head?

Until the third decade of the 17th century, it was widely believed the eyes sent an image to a *homunculus*, a 'little man', inside the head who made sense of what we saw. German astronomer Johannes Kepler wrote that everything we see appears before 'a magistrate sent by the soul'. This homunculus 'goes forth from the administrative chamber of the brain into the optic nerve and the retina to meet this image, as though descending to a lower court'.[25]

Not until 1637 did the one-time mercenary soldier and mathematician René Descartes lay to rest what philosopher Alva Noë has called 'the fallacy of the little man in the head'.[26]

In *La Dioptrique*, Descartes wrote: 'We must not hold that it is by means of this resemblance that the picture causes us to perceive the objects as if there were yet other eyes in our brain with which we could apprehend it.'[27] But, as Noë points out, 'if the retinal image does not function as a picture in producing vision, then it must function in some other way.'

To understand how, we must travel back to 1959 and a small, cramped laboratory at the Massachusetts Institute of Technology where Jerome 'Jerry' Lettvin was studying the visual system not of *Homo sapiens* but of frogs.

When Lettvin began his studies in the early 1950s, the retina was regarded as no more than a vast collection of light receptors that sent signals to the brain for interpretation. 'The assumption has always been that the eye mainly senses light, whose local distribution is transmitted to the brain in a kind of copy by a mosaic of impulses,' he says.[28]

Rather than accept that assumption, Lettvin and his colleagues attached electrodes to the frog's optic nerve and eavesdropped on the signals it sent.

After positioning an aluminium hemisphere in front of the frog's eyes, Lettvin moved objects attached to small magnets along the sphere's inner surface using a magnet on the outside. The frog could see these objects and follow them as they moved.

Analysing the optic nerve signals, they demonstrated the existence of 'feature detectors' – specific neurons in the retina that respond to features in the surroundings, such as edges, contrast, curvature, movement and changes in light levels. Lettvin and his colleagues also identified what they called 'bug detectors': clusters of cells that preferentially respond to 'small, dark objects that enter the visual field, stop, and then move around intermittently'.

Their conclusion?

'The eye speaks to the brain in a language already highly organised and interpreted instead of transmitting some more or less accurate copy of the distribution of light on the receptors.'

Although now understood as groundbreaking vision research, Lettvin's findings, published in his 1959 paper 'What the Frog's Eye Tells the Frog's Brain', were initially dismissed as nonsense. A fellow scientist at MIT branded him a liar; he was laughed off the stage at a major conference and threatened with having funding withdrawn 'if I didn't start behaving'. Only in recent years has the scientific world recognised the importance of his work and what it reveals about vision, as well as its role in how and why we experience blind spots.

Identical Signals – Different Perceptions

To the brain, electrical signals from your eyes, ears, tongue, skin, nose and every other sensory organ are identical. There is nothing to discriminate between signals from your eye and those from your big toe, or between those from your stomach and leg, your scalp and the soles of your feet – they all arrive in the brain as electrical impulses. But while the brain could reasonably plead that 'they all look alike to me', it obviously can distinguish between messages from your eyes, ears and other body parts. So how do our brains know whether the signals they receive represent the sight of a person playing the violin or the sound she is making?

An obvious answer is that different signals are processed by different regions of the brain. Signals from the eyes, for example, are mainly handled by the occipital lobes, at the back of the brain, while sound is processed primarily in the temporal lobes.

But, as Kevin O'Regan and Alva Noë point out, this is not the whole answer: 'Even if the size, the shape, the firing patterns, or the places where the neurons are localised in the cortex differ, this does not in itself confer them with any particular visual, olfactory, motor or other perceptual quality.'[29]

It is now recognised, as Lettvin's research established, that sight depends on the simultaneous action of thousands of specialised neurons that respond to a visual stimulus's strength, angles, shapes, edges and movements. These detectors work in parallel, each performing a specialised function.

What enables the brain to understand whether signals are being received from the eyes, ears, nose, mouth or skin are changes produced by the different muscles used to control the organ in question, such as those moving the eyes or turning the head. Imagine, for example, looking at the midpoint of a straight line, as shown in Figure 58, before shifting your gaze to a point some way above it.

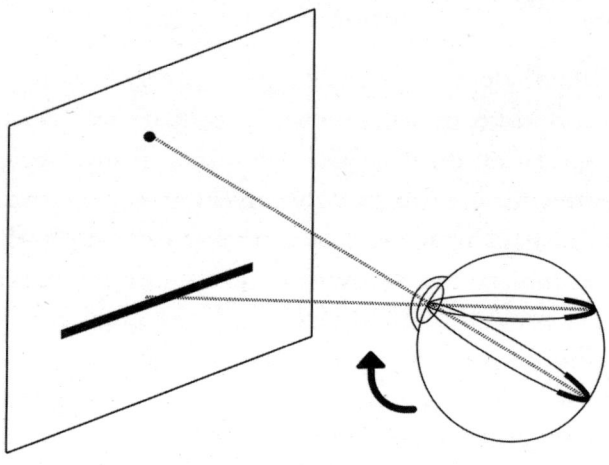

Figure 58

Like viewing it in a distorting mirror, the straight line dramatically changes as you move your eyes. Figure 59 shows how this movement causes it to display differently on your retina, sending a very different signal to the brain.

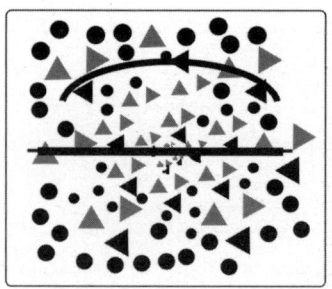 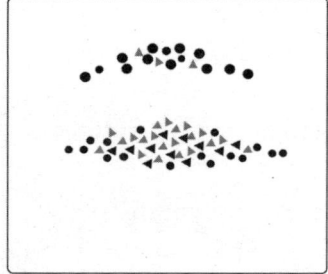

Figure 59

The diagram on the left depicts a flattened-out retina, with a straight line across the eyeball's equator, produced when looking directly at the line, and an arc produced when we look above it. Colour-sensitive cones are represented by triangles, and rods by discs. The corresponding cortical representation is shown on the right. The representation is fat in the middle when viewed straight on the line and tapers off to the ends. This changes into a thinner, banana-shaped region sampled mainly by rods when the eye moves upwards.

Regan and Noë put it this way: 'When the eyes rotate, the sensory stimulation on the retina shifts and distorts in a very particular way, determined by the size of the eye movement, the spherical shape of the retina, and the nature of the ocular optics ... as the eye moves, contours shift, and the curvature of lines changes.'[30]

Seeing, Remembering and Predicting
What does the photograph below show?
Your most likely answer will be a 'five-bar gate'.
But all you actually 'saw' were five horizontal, four vertical and four diagonal shapes, which, by activating the brain's parallel, horizontal and diagonal feature detectors, triggered a memory of similar patterns stored in memory – in this case, of a five-bar gate. Suppose you had never learned that pattern's linguistic label. Then, you would have seen the same object as

Figure 60

a meaningless jumble of shapes, lines and forms. Consider, for example, the following ideogram.

五槓門

To a Chinese speaker, this will instantly be recognisable as *Wǔ gāng Mén* (five-bar gate). But if you do not understand the language, the horizontal and vertical lines will have no meaning for you. Making sense of what you see is as much about recollection as perception. Visual memories are rarely consciously experienced. Unlike occasions when we must actively use our memory, for example when recalling facts or figures, overlearned skills such as those of seeing are accessed unconsciously and, as a result, without our having any knowledge of utilising them.

Even a minimal amount of visual information is often sufficient to establish an accurate memory of what you are looking at. For example, the arrangement of horizontal, vertical and diagonal lines in Figure 61.

Partially obscured by greenery, it could be a gate or a fence. Yet the pattern remains sufficiently distinctive for most people to identify it as a gate. When we presented this question on TikTok, 48 per cent correctly identified it as a gate and 30 per cent as a fence, with just 7 per cent being unable to decide either way. The remaining 15 per cent gave answers that were

Figure 61

neither a gate nor a fence. Had they seen the obscured gate in real life, it's far less likely anyone would have been confused. This is due to another, seldom fully appreciated, feature of our visual system. Having two eyes endows us with X-ray-like powers![31]

Test this for yourself by splaying the fingers of each hand and holding them up in front of your face. Now note how much you can see with either your left or right eye closed and compare it with how much you can see when both are open. Combining the two images provided through binocular vision enables us to see around and partially through objects in our surroundings – an ability with obvious survival advantages for our hunter-gatherer ancestors. Spotting prey or predators through the tall grass or dense foliage would confer a decisive evolutionary advantage.

Another little-appreciated aspect of vision is that, due to the time it takes for an image to be perceived by the eye and processed by the brain – around one-tenth of a second – we never see the world as it is but as it was a fraction of a second earlier.

Imagine trying to catch a cricket ball travelling at 160km/h or striking a tennis ball at 120km/h. In the time it takes for information about its position to reach your brain, the cricket ball will have travelled some 4.4m, and the tennis ball around 3.4m. This means you see the ball not where it is but where it was! How can anyone ever catch or strike such a fast-moving object?

To answer this question, cognitive neuropsychologist Tessel Blom and her colleagues recorded volunteers' brain activity as they followed fast-moving objects. 'We suspected the brain might solve its delay problem by making predictions,' she explains. 'If that were true, we reasoned … it would take time for the brain to "discover" that the object was gone. The brain would briefly "see" the object beyond the point where it disappeared.'[32]

Their recordings confirmed this. When a moving object suddenly disappeared from view, the volunteers' brains continued to 'see' it because that is what they expected to see. We perceive the present only by predicting the future and making it a reality by acting on those predictions.

'As strange as it sounds, when your behaviour is involved, your predictions precede sensation, but they also determine sensation,' says neuroscientist Jeff Hawkins. 'Thinking, predicting, and doing are all part of the same unfolding of sequences moving down the cortical hierarchy.[33]

A World of Our Own

Every animal on this planet inhabits its own perceptual world, or *Umwelt* – the German word for 'environment', first used in this context by biologist Jakob von Uexküll in 1939.[34] Born in 1864 as heir to an extensive estate in the Governorate of Estonia, Jakob enjoyed a life of leisure and luxury for more than half a century. This ended abruptly with the Russian Revolution in 1917. After the Bolshevik Party confiscated his fortune and estate, a now-destitute Jakob urgently needed a job. With an excellent brain and a strong interest in biology, he obtained a position at the University of Hamburg, where he founded the Institut für Umweltforschung (Institute for Environmental Research). Jakob believed that every living thing occupied its own 'bespoke sliver of reality'.

What this reality is depends on how any particular creature makes sense of the world. For example, a tick's *Umwelt* is limited to three sensations: light, odour and body temperature. This eyeless animal, Jakob pointed out, could find its way to the top of a tall blade of grass or a bush using its general sensitivity to light as a guide. Lying in ambush, this otherwise blind and deaf creature would await the arrival of a warm-blooded animal, such as a sheep or goat. Then, employing its only other sense, smell, to detect the presence of butyric acid emanating from glands in the animal's skin, it would abandon the grass or bush. 'If she is fortunate enough to fall on something warm (which she perceives using an organ sensible to a precise temperature),' von Uexküll wrote, 'then she has attained her prey, the warm-blooded animal, and after that needs only the help of her sense of touch to find the least hairy spot possible and embed herself up to her head in the cutaneous tissue of her prey. She can now slowly suck up a stream of warm blood.'[35]

The *Umwelt* of sharks and platypuses include sensing electric fields, while rattlesnakes detect infrared radiation, and vampire bats perceive ultraviolet light. For an albatross, oceans become a landscape of odours revealing the presence of food.

In his classic 1974 essay 'What Is It Like to Be a Bat?', philosopher Thomas Nagel wrote about the insurmountable difficulties of describing or understanding the conscious experiences of other people and species. He explained how bats use echolocation to precisely discriminate distance, size, shape, motion and texture like sighted creatures do with their eyes. 'But bat sonar, though a form of perception, is not similar in its operation to any sense we possess, and there is no reason to suppose that it is subjectively like anything we can experience or imagine.'[36]

Indeed, even within the same species, we are unlikely to see the world exactly as others do. We were flying to a Berlin conference on forensic science accompanied by George, a Scotland Yard detective, and Barry, a medical doctor. When our flight was delayed by fog, Barry suggested we pass the time by playing a game he called 'spotting the lesions': he would diagnose other passengers by looking for signs of disease and disability. George, meanwhile, would see how many villains he could spot, while we watched for people falling victim to blind spots.

By the time our flight was called, Barry had identified 20 physically sick passengers and George had detected six potential villains. We had spotted a dozen failing to notice directional signs, bumping into other people, and forgetting pieces of luggage; helped by the fact that inattentional blind spots are likely at airports, due to higher-than-usual anxiety levels.

Even though we were in the same environment, we saw it subtly differently because of the different purposes directing our attention. With Barry, it was pathology; with George, it was criminality; and for us, blind spots. Most of the time we are less consciously aware of our purposes in how we are viewing the world, but such guiding principles are always present.

Jakob von Uexküll's *Umwelt* theory, one of the most elegant and insightful in biology, tells us that the all-encompassing nature of our subjective experience is an illusion. We are only ever aware of a fraction of what there is to sense. Our senses constrain us, creating a permanent divide between our *Umwelt* and that of every other animal. 'Our brain does not see but constructs reality,' explains psychologist Christoph Witzel, 'It infers an outside world from ambiguous input.'[37]

So far, we have examined the different forms blind spots can take and explored some of the science behind how we see and make sense of the world around us. In the following chapters, we examine the role played by different forms of psychological blindness in everyday life, from recognising faces and going shopping to running a business – all activities where outside agents can and do exploit our visual vulnerabilities for their own purposes.

CHAPTER EIGHT
Why We Never Forget a Face

Once someone's face has been stored in our long-term memory, it will, barring traumatic brain damage or neurodegenerative diseases like Alzheimer's, remain there for the rest of our lives.

Facial recognition is essential for social survival, so our inability to consciously forget a face is unsurprising. As vision psychologist Vicky Bruce points out: 'The human face provides a bewildering variety of important social signals which can be detected and interpreted, very often correctly, by another human. A face tells us if its bearer is old or young, male or female, sad or happy, whether they are attracted to us or repulsed by us, interested in what we have to say or bored and anxious to depart.'[1]

With little effort, we store and retrieve thousands of faces – although not always the names that go with them – of our family and friends; neighbours and colleagues; strangers passed on the street, met on a train, or noticed in a restaurant. We recognise them in the flesh, in a photograph, or in a home movie, at a distance, under poor light, and despite changes due to ageing, growing a beard, changing their hairstyle or even having cosmetic surgery.[2]

While you may believe every detail matters when recognising someone's face, studies have shown that your brain uses a far coarser resolution. Matthias Keil found that even with a low-resolution image (30 x 40 pixels) of only the mouth and eyes, people could identify familiar faces almost instantly.[3]

In a social media poll we ran, 95 per cent of people instantly recognised this famous face when shown only the eyes and mouth. If you are having difficulty, you'll find the answer at the end of this chapter.

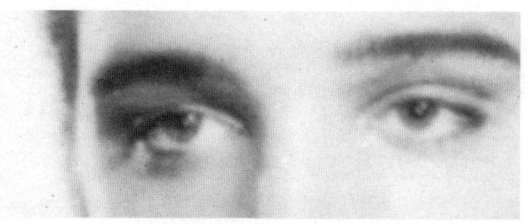
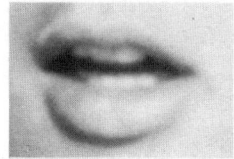

Figure 62

Recognising familiar faces from such meagre visual clues is a skill we learn early on. Possibly long before birth.

An intriguing series of studies conducted by cognitive neuroscientist Vincent Reid and his colleagues showed that babies react strongly to face-like stimuli long before birth. Using lasers, they shone three dots of light into the wombs of pregnant women. As shown in Figure 63, these formed two types of triangle. One had a point at the top and the other at the bottom. The researchers argued that the latter was more face-like since it mimicked our broad forehead and narrower chin.[4]

In the figure, the upper images represent the face-like pattern and the lower images the non-face-like pattern. The images on the left are how the pattern appeared when projected and those on the right show how it would have appeared to the foetus in the womb.

Using ultrasound to monitor the babies' responses, Reid and his team discovered that while unborn infants turned their heads to follow the face-like pattern, they made no such movements for the non-face-like shape. Psychologist Dmitry Kobylkov and his colleagues confirmed this innate ability to

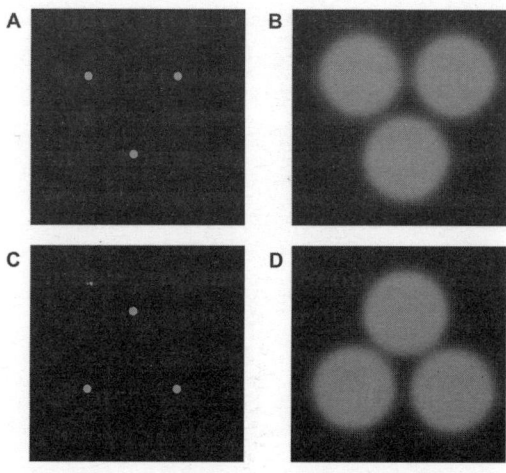

Figure 63

detect face-like patterns, working not with babies but with week-old chicks.[5]

Kobylkova's team identified neurons in the caudolateral nidopallium, the avian equivalent of the prefrontal cortex in mammals, in the chicks that, despite never having been exposed to faces, responded to a face-like stimulus composed of three dots judged to resemble two eyes and a beak (or a mouth).

Studies such as these have fuelled a debate on the role of nature versus nurture in developing face-detection mechanisms. Three theories exist about how and why faces fascinate young animals.

The first, illustrated by the examples above, proposes that animals are born with a 'facial template' enabling them to detect the inverted-triangle arrangement of eyes, nose and mouth. Human studies such as that by Reid suggest we have an innate mental ability to identify the configuration of human faces – that is, the position and distances apart of the various facial features.[6] One advocate of this theory, Jennifer Richler, notes that: 'Because faces are made from common features (eyes, nose, mouth, etc.) arranged in the same general configuration, subtle differences in spatial relations between face features being encoded [are] beneficial for successful recognition of a given face.'[7]

This sensitivity also explains, for example, why sometimes we see faces in the clouds or spots on the walls – the phenomenon known as pareidolia that we encountered earlier.

Figure 64 *Figure 65*

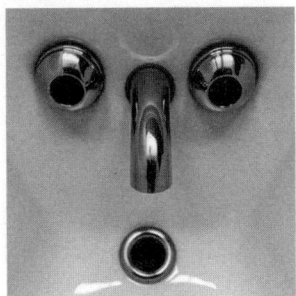

Figure 66

As with the scorch-marked tortilla described in Chapter 5, the very different objects pictured in Figures 64–66 – an alarm clock, a tap and a US mains socket – are seen as having 'face-like' qualities.

An artist who recognised our ability to perceive faces in the least likely places was the 16th-century Italian painter Giuseppe Arcimboldo. He became famous for portraits composed entirely of fruit and vegetables, such as one in

Figure 67

which the Holy Roman Emperor Rudolf II is depicted as Vertumnus, the Roman god of the seasons. Looking at the picture from a distance, it appears conventional. Come in closer, and the face looks very different.

Yet other researchers have cast doubt on this 'configuration theory'. For example, Mike Burton and his colleagues make the point that: 'If recognition relies on "subtle differences in spatial relations between features", then it follows that disrupting these spatial relations should harm recognition. However, this is not borne out by the evidence. Instead, recognition of familiar faces appears remarkably robust under a range of deformations.'[8]

The second theory then proposes that we perceive faces holistically rather than seeing separate features like the mouth, the nose, the eyes, etc. Studies have found that people more efficiently and accurately recognise a feature such as a nose when it is seen as part of a whole face rather than in isolation.⁹

Indeed, one of the critical aspects of holistic vision is our ability to recognise objects based on very little information. If, for example, a glimpsed face barely emerges from the shadows, it may be instantly recognisable even though only a few patches of dark and light register on the retina. Any missing details are inferred and the object is perceived based only on a few visual 'hints' in a process of visual completion based on prior perceptual knowledge called 'perceptual closure'.[10]

The third theory proposes that if no inborn blueprint exists for recognising faces, one develops through learning and experience.

After becoming a father, brain scientist Pawan Sinha decided to find out where newborns direct their gaze, using his infant as a guinea pig. He recorded his child's worldview with a miniature, headband-mounted video camera and calculated how much time was devoted to studying faces.

'Given that the parents hold the newborn much of the time while awake, faces end up being the most prevalent object in the inputs,' reports Sinha. 'Furthermore, the infant's poor acuity significantly degrades images of objects beyond a few feet, adding to the close-up faces' salience. These aspects of the baby's experience make it feasible for an unsupervised learning system to rapidly acquire a rudimentary face concept, even without any innately specified face templates.'[11]

How Babies Communicate Before They Can Speak

In another study, cameras simultaneously recorded the faces of mothers and their newborn babies as they interacted. The pictures on the next page show one mother using eye contact and expression to focus her baby's attention on her face.

Figure 68

The dialogue commences with the mother Gina calling out to six-month-old Thomas. Hearing a familiar voice, the little boy focuses intently on his mother's face, starting the 'orientation' phase of their 'conversation' (Figure 68).

Figure 69

Gina smiles in greeting, and Thomas responds, looking pleased and alert (Figure 69). He holds the mother's face in a steady gaze and carefully follows her movements.

During the 'play dialogue' stage of their nonverbal 'conversation', Thomas attempts to mimic his mother's lip movements.

Studies like this suggest our ability to understand the meaning of facial expressions is part nature and part nurture. However, this does not apply to *forming* those expressions that seem entirely innate and require no learning. Our facial muscles can produce more than 40 independent actions, resulting in many possible expressions. There is strong evidence that a small number of these are innately and universally produced in response to specific emotions.[12]

A study of expressions in congenitally blind athletes by psychologists David Matsumoto and Bob Willingham found no differences in expression between them and sighted athletes. 'These findings provide compelling evidence that the production of spontaneous facial expressions of emotion is not dependent on observational learning,' says Matsumoto.[13]

Research also indicates that our brains take in more information than just facial movements when understanding facial expressions, with gender and culture influencing this process. The greater our exposure to a wide range of facial expressions, the more likely we will recognise them accurately and respond appropriately. We can, for example, learn to distinguish a genuine from a polite smile by paying attention to the eyes rather than solely the lips. One of the telltale signs of a sincere smile is the fine lines – 'crow's feet', or lateral canthal lines – that form on the eye's outer corners. Genuine smiles involve the muscles that lift the corner of the mouth upward and laterally *and* those around the eyes (the orbicularis oculi muscles). Smiles produced to be socially gracious or when in uncomfortable situations use only muscles in the lower part of the face.[14]

The Two Faces of Margaret Thatcher

While an inborn facial template and early learning play vital roles in recognising and remembering faces, however, the arrangement of features and the spatial difference between them are only part of the story. This was demonstrated by a landmark study of the 1980s that unexpectedly became famous as satire despite only being intended to offer scientific insight. Psychologist Peter Thomson made minor changes to the face of then British Prime Minister Margaret Thatcher, then turned the image upside down before asking subjects whether they could spot the changes.[15]

We have updated his classic study using the *Mona Lisa* to display the same effect. Look at the two images on the next page and see whether you can detect any significant differences.

As in Peter Thomson's Margaret Thatcher study, while no one viewing this world-famous painting has difficulty recognising it as the *Mona Lisa*, they fail to notice differences between the two images – until they turn them the right way up, and the differences become apparent. To view this transformation, either turn the book upside down, go to lewisandleyser.com/thatcher-effect or scan the QR code.

Due to an interaction between face geometry and the direction of shading and shadowing across its surface, we are used to seeing faces as patterns of light and shade with shadows below the eyebrows when a face is lit from above. Over half a century ago, cognitive psychologist Craig Mooney demonstrated this by creating 50 different faces depicting people of all ages, only defined by patches of black and white.[16]

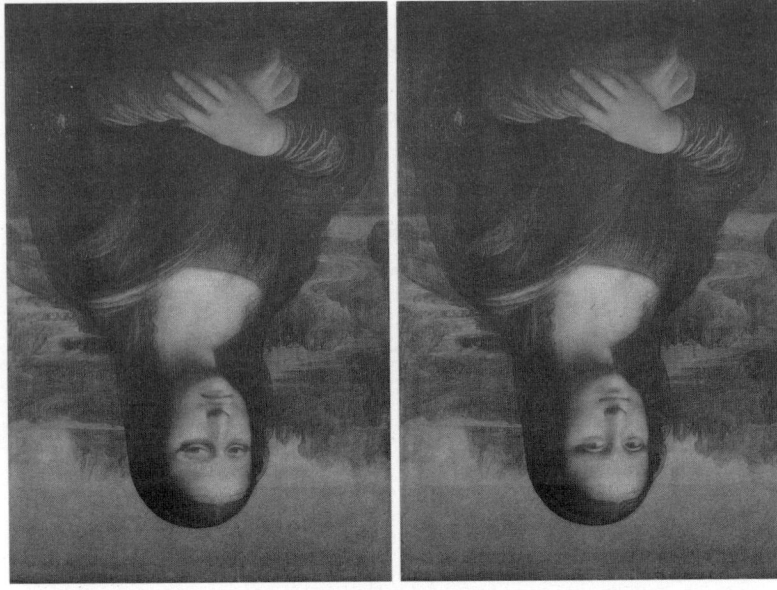

Figure 70

Mooney faces such as in Figure 71 are now an essential tool in studying human vision: they are highly degraded versions of facial images that humans can still perceive as human faces. While they may take a few seconds to recognise, once perceived, it is difficult for a normally sighted human observer not to see the face from that moment onward.[17]

Figure 71

How Unseen Expressions Can Change Behaviour

We can recognise faces in as little as 50ms. This is too brief a time to become consciously aware of what we're seeing but sufficient to identify a familiar face and even to be affected by its expression. This was demonstrated in an intriguing experiment conducted by psychologist Mark Baldwin and his colleagues, in which a group of Catholic women were asked to read a story describing a woman's sexual dream. Having read the passage, they sat opposite a display and saw five brief light flashes. They did not know they had also been subliminally exposed to a picture of either the Pope or a censorious expression on an unknown man.

The participants then completed a questionnaire to evaluate their religious commitment. Their answers were compared with self-assessments they had made before being shown the images. The women exposed to the unfamiliar face showed no change in religious commitment. Those exposed to a brief flash of the Pope's disapproving features gave themselves lower ratings despite never consciously registering the face.[18]

How Brains Recognise Faces

In 2011, neuroscientist Josef Parvizi used deep brain stimulation to conduct groundbreaking studies into human perception. Sometimes used with epileptic patients, deep brain stimulation involves drilling holes in the skull, implanting electrodes and stimulating the brain with small amounts of electric current. By doing so, doctors can prevent seizures from occurring.

However, Parvizi found he could also distort how epileptic patients saw faces by stimulating a specific brain region known as the fusiform face area (FFA). One man looked steadily at the face of Dr Parvizi's female colleague as a small current coursed through this area of his brain. He told them in astonishment: 'The bottom of her face metamorphosed. Kind of stretched up to give her a different look. Her nose got misshapen a little bit. Her lips just changed. It wasn't pretty! She turned into someone else, like someone I've seen before but a different person in my memory. Only her face changed. Everything else remained the same.'[19]

When another patient's FFA was stimulated in a related study, he reported seeing faces everywhere he looked so long as the electric current flowed. If he looked at an egg, a pen, a magazine or a flower, he saw these objects with a superimposed human face.

This research helped confirm that specific brain regions respond far more strongly to faces than anything else – a finding earlier made by Nancy Kanwisher and colleagues at the end of the 1990s.[20] Then, Kanwisher had asked participants in her study to view faces or objects, including houses, telephones and animals. She then used functional magnetic resonance imaging (fMRI) to see which brain areas became most active when doing so, allowing her to identify what she called 'functional localisers'. The regions showing the most significant activity when viewing faces were given the FFA designation encountered above (the first 'F', fusiform, means 'spindle-shaped'). These were located primarily, though not exclusively, on the right side of the brain. For comparison, subjects were also shown objects other than faces, whereupon the region showing the most activity was further back and slightly closer to the skull – a region known as the visual or occipital cortex.

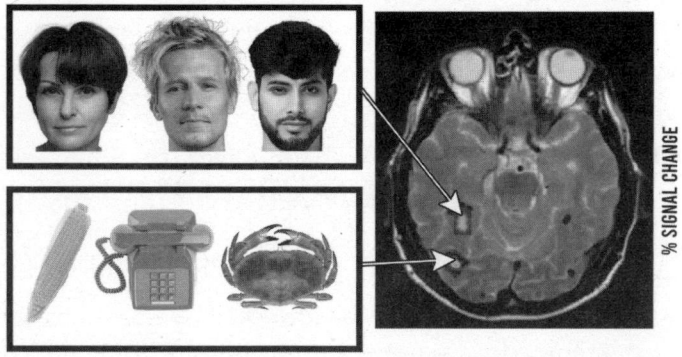

Figure 72

Damaged Brains Fail to Recognise Familiar Faces

The theory that we can only recognise faces using the fusiform facial area is supported by the fact that if that area is damaged, people cannot recognise even familiar faces. This condition is known as prosopagnosia (from the Greek *prósōpon*, for 'face', and *agnōsía*, meaning 'not knowing'), and was first identified at the end of the 19th century.[21]

The late physician and neurologist Oliver Sacks, himself a prosopagnosia sufferer, tells the story of a patient eating in a restaurant who complained to the waiter that a stranger sitting opposite was constantly staring at him. 'Every time I look up, the man looks straight back at me.' The baffled waiter pointed out that no one was sitting opposite. The man was looking at himself in a large mirror. Damage to the unfortunate man's FFA prevented him from recognising his own face.

Our seemingly effortless facial recognition has a Jekyll and Hyde aspect. Because information is skipped over in the name of speed and efficiency, when studying someone's face far more remains unseen than seen.

You may think mistaking the face of one stranger for another is unsurprising, especially after seeing them only briefly. After all, when meeting someone for the first time, we may be distracted by wondering: do they look friendly, relaxed, bland, animated, expressionless, happy or sad? Or even: do they find me attractive? Do I find them appealing?

Preoccupations such as these make it difficult for many to remember that person's face a few moments after being introduced. Yet there are also occasions when we may fail to recognise even a *familiar* face, which can happen when they either appear in an unexpected context or when we are emotionally aroused. While shooting *Give My Regards To Broad Street* in 1984, Paul McCartney started busking outside London's Leicester Square underground station. Despite being one of the world's best-known musicians, passers-by ignored him: 'I was standing there plunking chords, doing this silly version of "Yesterday", and no one noticed it was me,' he recalls.

In a big city, people generally avoid eye contact with strangers, so their failure to recognise even someone as famous as Paul McCartney may seem reasonable. Under the same circumstances, however, it would seem far less likely that people would fail to identify the face of a near relative or close friend. To put this to the test, psychologists Sarah Laurence, Jordyn Eyre and Ailsa Strathie showed their student subjects a video depicting a man and a woman. While the man was unknown to them, the woman was a psychology lecturer who had taught most of them numerous times. Yet after watching the video, none of the students recognised her. Even after viewing it several times, less than 10 per cent of them did so.[22]

A cynic might say it is hardly surprising that students fail to recognise their lecturers since they spend much of their lecture time reading their WhatsApp messages or scrolling social media. But no one would argue that failing to identify members of one's own family is due to a lack of familiarity. Yet this scenario too has been tested, by Donald Thomson in 1986. He asked a London-based Australian student to stand outside her parents' hotel when they were visiting the city. Unaware their daughter was also in London, they walked past her with no signs of recognition.[23]

Experiments such as these illustrate the importance of context when recognising faces. Expectation blind spots can cause us to fail to see what we don't expect to see.

Our emotional arousal can also play a major role in facial recognition. Although we may easily recognise a familiar face when feeling calm and relaxed, our ability to do so quickly deteriorates when we are afraid or angry, as we saw with Ottilie Meissonier, whose mistaken identification of businessman Adolf Beck we described in Chapter 3. Furious at being defrauded and betrayed, she misidentified him, with disastrous results.[24]

Indeed, so fearful are assault victims when being attacked, they typically have great difficulty in identifying their assailants. This was dramatically demonstrated by psychologists Brian Clifford and Clive Hollin, who produced a series of videos, three of a violent incident and three of a similar but non-violent one.[25] Each opened with a shot of the same woman walking towards the camera. In the violent sequences, a man ran into the frame – allowing viewers a clear profile view – forced her against a wall and grabbed her handbag. He fled the scene – providing a full-face view – leaving the victim sobbing.

Other videos depicted the same robbery, but on these occasions, the bag snatcher was accompanied by either two or four different men, who also menaced the woman. In the non-violent versions, the same man, accompanied by two or four others, merely approached the woman to ask for directions.

Subjects were randomly assigned to watch one of these videos before being asked to identify the men. They were shown an array of photographs and informed that one was of the man who, depending on which video they watched, had snatched the handbag or sought directions. Clifford and Hollin found worse identification for the violent than the non-violent scenario, with this impact being compounded the greater the number of men involved. When five men were involved in the violent incident, 90 per cent could not identify the handbag snatcher, even though an incident as dramatic and novel as a bag snatch is likely to produce what is known as a 'flashbulb' memory, whereby witnessing a shocking and stressful incident generates a vivid and enduring memory, though of the event itself, rather than the details.[26]

They Say Love is Blind – It Is

The idea of eyes meeting 'across a crowded room' as two people instantly find a lifelong partner has been a prominent theme in arts and literature for at least three thousand years. It sounds wonderfully romantic, but is there any truth to it?

On the one hand, studies have shown that 60 per cent claim to have experienced love at first sight; perhaps the same has happened to you.[27] On the other hand, statistics show that in the UK, seven out of ten first marriages and four out of ten marriages overall end in divorce.[28] According to the American Psychological Association, approximately 40–50 per cent of first marriages in the USA end in divorce, as do 60–67 per cent of second marriages.[29] What initially seemed like love at first sight all too often ends in betrayal, rejection and bitter recriminations.

Indeed, 'the moment of love, at first sight, does not seem to be marked by a high passion for a person and does not seem to involve feelings of love at all but a readiness to experience them at best,' says psychologist Florian Zsok.[30]

This creates an emotional attachment during which lovers 'communicate their affection via a copulatory gaze, in which the individual gazes into the other person's eyes for many seconds'.[31]

Love at first sight is often a one-sided affair. The intense initial experience of one participant can shape the other into believing they also experience it. Both will radically change their behaviour due to that overwhelming feeling, going to great lengths to make themselves as desirable as possible to the other. They adjust how they dress, what they say, and their behaviour. They are often willing to discard long-term dreams, abandon much-desired ambitions, and sacrifice deeply held beliefs to win the other's love.[32] Lifelong friends will be discarded, family members be sidelined and even common sense betrayed to maintain the relationship. Men, although not exclusively, often misinterpret politeness or friendliness from another as indicating strong sexual interest. This can lead to repeated attempts to develop an intimate relationship, even though such efforts will be fruitless.[33]

Whether or not love at first sight exists, it's important to understand how we perceive those we are attracted to. To explore the truth about physical attraction, researchers used eye-tracking to record the gaze patterns of young (average age 20) heterosexual men and women as they were shown images of stereotypically attractive members of the opposite sex. These enabled them to identify which parts of the anatomy went unseen and which were examined most closely.[34]

Figure 73 shows how young men viewed a female (left) and male (right) model. Figure 74 shows how young women viewed the same images. The male gaze for the photo of the female model was concentrated almost exclusively on the breasts and genitals. At no time did they pay any attention to her face. When young women saw the same image, their attention on viewing the female was split more or less equally between her face, breasts, abdomen and groin. On the male, much attention was paid to his facial features, especially the eyes and mouth, with focus also on his chest, stomach and arms, with the only interest below the waist being his knee.

Evolutionarily, these male and female gaze patterns make perfect sense. The concept of sexual selection – whereby a mate is chosen based on the best chance of reproductive success – was introduced by Charles Darwin in the 19th century as part of his theory of natural selection. He claimed the 'sexual struggle is of two kinds'. The first 'between individuals of the same sex ... in order to drive away or kill their rivals', and the second 'between individuals of the same sex, in order to excite or charm those of the opposite sex'.[35]

The latter is argued to be done through the display of traits that are evaluated by the opposite sex to determine the health and fertility of a potential mate, explaining the gaze patterns of both the men and women in our study. Men focused on areas that likely signal the ability to support gestation and

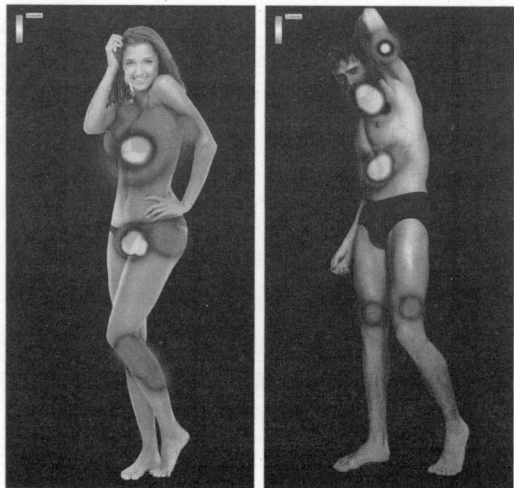

Figure 73

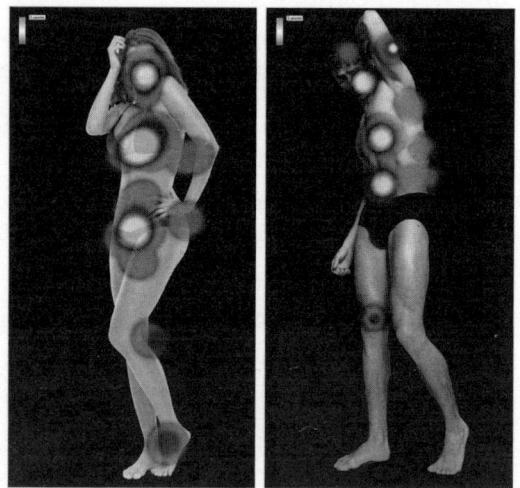

Figure 74

lactation for offspring, while women focused on areas that might give overall indicators of health and therefore fertility. This has been corroborated by other studies that have also found women focus on facial shape, musculature and height when viewing pictures of potential mates.[36]

However unlike men, women have also been shown to have a preference for partners with not just preferred physical traits, but also preferred social traits related to resource accrual.[37]

To understand why, we have to go to the gamete level. While only a single ovum and single sperm is required to reproduce, men produce sperm with each ejaculate, whereas women have a finite number of ova they are born with. This asymmetry leads to what has been termed 'Bateman's principle', whereby access to resources is more important for women as they cannot reproduce as frequently. This has led to parental investment theory,[38] whereby women are argued to have a higher obligatory investment in offspring compared to men and so be more selective in choosing reproductive partners who can provide for any future offspring.[39]

Indeed, studies have found that women emphasise factors such as ambition and career-orientation — though only when considering long-term rather than short-term partners.[40]

In 1993, the nature and extent of mate selection differences was researched by psychologist Michael Wiederman based on replies to over a thousand personal advertisements.

He reported men were more likely than women to mention financial resources, honesty and sincerity when wooing a potential mate. In return men looked for an attractive face and appealing body shape, typically in a woman younger than themselves, a trend that became more pronounced the older they were. They were also more likely than women to make explicit requests for a sexual rather than a purely companionable relationship.[41]

Women, by contrast, were more likely to seek a man who, while having an attractive face and body, combined financial resources with honesty and sincerity. They were also more likely than men to seek companionship and only a sexual relationship after friendship had been established.

What Makes a Person Attractive?

Physical attractiveness has consistently been shown to be a strong predictor of attraction and partner choice.[42] This holds true across cultures, sexes and whether involving short-term or long-term partner preferences.[43] Immediate physical attraction is especially important when it comes to love at first sight.[44] Another interesting finding from studies on attractiveness is that people who are viewed as more attractive are also perceived as more friendly, healthy, intelligent, and trustworthy.[45] This mutual positivity helps couples become more intimate, attentive, and caring towards one another – at least as long as the romance lasts! They are also more likely to be hired for jobs than less attractive individuals, granted bail if accused of a crime, believed when giving evidence and less likely to be convicted by a jury.[46]

There is a widely held belief that whether or not we find someone attractive has been learned via culturally presented ideals, especially the media.[47] If this were true, beauty could mean different things in different times and places. Yet while individual and cross-cultural differences exist, there's not as much difference as expected. Many studies have found a high degree of agreement between individuals within and between cultures when judging faces of varying ethnic backgrounds, suggesting that people everywhere use the same, or at least similar, criteria in their judgements.

The speed with which we process facial information enables us to evaluate a potential sexual partner in about a tenth of a second. In that brief moment, our decision will largely depend on three physical features: similarity, symmetry and, to a lesser extent, eye colour.

Similarity
In a 1999 study, evolutionary psychologist Ian Penton-Voak and his colleagues photographed 52 female and 21 male students. They then morphed these photos to create

Figure 75

opposite-sex images with facial features on a spectrum of similarity to the original.

We have recreated this study, with Figure 75 showing the authors in the top two positions on the left, transformed into similar and dissimilar female versions of themselves. The lower two images show two women (left) transformed into similar and dissimilar men.

Participants, who had no idea their faces had been used to generate the opposite-sex image, were shown these photos and asked to adjust the face's appearance using a computer mouse so that the features more or less closely matched their own. They could select a face between 100 per cent similar and 100 per cent dissimilar. The researchers found that, up to a certain point, the greater the similarity to their own, the more strongly they were attracted to it, with attractiveness declining sharply the less similar it was to their own. 'Highly unusual faces are likely to receive low ratings of attractiveness,' says Penton-Voak. 'Subjects are not attracted to faces that look very different from themselves, but neither is anyone else. This means that average face shapes will have moderate levels of similarity to most subjects and will be rated relatively highly for attractiveness.' Like Narcissus, the handsome young man in Greek mythology, we are in love with our own appearance.[48]

This suggests that, while the old saying 'like attracts like' is accurate up to a point, when choosing a romantic partner, we are not looking for a mirror image but a person who appears very like or somewhat similar to ourselves.

Symmetry

Facial symmetry, the second element of mutual attraction, signals a would-be partner's health and genetic viability. Ideally, the same genes should be activated in the left and right sides of our faces, resulting in perfect bilateral symmetry.[49] In reality, though, no human face is precisely symmetrical. Tiny fluctuations in gene expression and cellular activity lead to detectable differences between the two halves of every face. Study faces carefully, and you often see one eye is very slightly larger and a little higher than the other. The nostrils, too, may be asymmetrical in size and shape, as are the height and size of the ears. Typically, there is a slightly wider separation between the edge of the mouth and the eye on one side than the other. All these differences, however minor, contribute to the face's asymmetry and greatly influence how attractive others will find it.

Because we tend to fix our gaze on another's eyes and mouth, the angle between them provides a valuable symmetry indicator (see Figure 76 on the next page). According to a study by psychologist Daphne Maurer and her colleagues, faces with smaller eye–mouth–eye (EME) angles are perceived as more attractive, especially for men.[50]

Male facial symmetry is associated with lower levels of the steroid hormone cortisol; such low levels are associated with better health and higher levels of the sex hormone testosterone, which is a predictor of fertility. The research suggests, then, that the symmetry in a man's face can predict reproductive potential.[51] This may be one reason why symmetry has such a significant effect on how attractive we find others. But it may not be the only one.

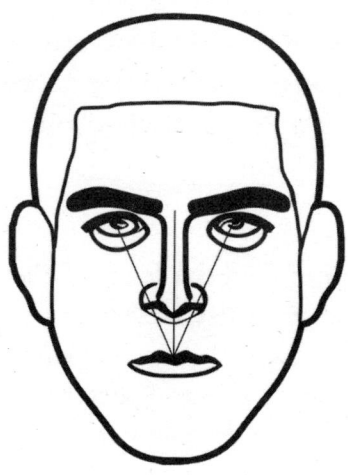

Figure 76

The evolutionary advantage theory holds that anything less than perfect symmetry indicates some dysfunction, however small. Micro-asymmetry, differences too small to be seen with the naked eye, although which we may be subconsciously aware of, are due to a gene being over- or under-expressed, expressed too early or too late, or in the wrong place. More significant and easily seen differences may indicate potential handicaps to the individual's and their offspring's success. These include DNA damage, infections, inflammation, allergic reactions, injuries, mutations, malnourishment, parasites, and genetic and metabolic diseases. Facial symmetry influences our choice not only of mates but also of friends and acquaintances.[52]

This does not mean that only men and women with symmetrical faces can ever expect to find love, beget children, and enjoy long-term relationships. Many other factors, both physical and psychological, play an increasingly important role as a relationship develops and matures. While facial symmetry is important for both sexes, studies have found bodily symmetry is also an important feature men look for in women, whereas social status and wealth have been shown to be of greater importance than physical looks or health in men.

However, facial attraction does play a crucial role in determining whether a relationship will begin. Even Romeo and Juliet might never have met without an initial spark of mutual interest.

Eye Colour

Human eyes are unique in the relatively large amount of the white casing of the eye (the sclera) they reveal in comparison to other primates. One study examining the eyes of 88 different primate species found that humans not only had the largest areas of exposed sclera and the most horizontally elongated eyes but were the only ones whose sclera was white. In most primates, it matches the skin colour around the eyes.

The researchers suggested that having the direction of gaze camouflaged may be adaptive for many primates since direct eye contact often provokes attacks, whereas in humans, enhancing gaze signals through shape and eye colour may aid communication and cooperation between individuals, particularly in social groups.[53] Eye colour depends on the ratio of dark brown eumelanin to reddish pheomelanin and how these are distributed. Blue eyes are primarily due to a lower amount of brown eumelanin rather than pheomelanin and tend to have very little melanin, causing light to scatter and appear blue.[54]

There have been interesting links found between attractiveness and eye colour, though surprisingly only for blue-eyed men. Neuropsychologist Bruno Laeng and his colleagues showed pictures of blue- and brown-eyed men and women to members of the opposite sex. They found men with blue eyes rated blue-eyed women as more attractive than women with brown eyes. However, there were no eye-colour preferences for men with brown eyes, women of any eye colour, or non-heterosexual participants of either sex.[55]

The researchers reported similar results when participants rated the attractiveness of pictures of young men and women whose eye colour changed but whose features otherwise remained the same. Among female participants there was no difference in the attractiveness ratings given to men with blue compared with brown eyes. In contrast, blue-eyed men rated blue-eyed women as more attractive than brown-eyed ones.

The researchers concluded that when a heterosexual blue-eyed man seeks a mate, this preference allows him to have 'paternity assurance'. As a recessive gene causes blue eyes, eye colour provides a prominent and easily visible clue to the child's heredity. This has given rise to the widely believed genetic myth, first mentioned in an academic paper over a century ago, that a single gene determines blue eyes, meaning blue-eyed couples only have blue-eyed children.*[56]

Our journey through the fascinating world of facial recognition has revealed the intricate and robust mechanisms that underlie this fundamental human ability. However, our facial recognition abilities, while impressive, are not infallible. Expectation blind spots can cause us to overlook even familiar faces in unexpected contexts, and emotional

---

*To avoid any blue-eyed male readers with brown-, hazel- or grey-eyed children freaking out on reading this, we should add that eye colour is associated with not one but several genes, and the interactions between them. It is possible, if rare, for a pair of blue-eyed parents to have brown-eyed children.

states, particularly fear or anger, can significantly impair our ability to recognise and remember faces accurately. These limitations have profound implications, especially in high-stakes situations like eyewitness testimonies in criminal cases.[57]

Understanding the mechanisms and limitations of facial recognition can help us in numerous ways. In security and law enforcement, it can inform better practices for eyewitness identification. In social interactions, it can help us be more aware of our biases and the judgements we may make based on facial features. In personal relationships, it can provide insights into the initial spark of attraction and the development of long-term bonds.

As technology increasingly attempts to replicate and even surpass human facial recognition abilities, understanding our cognitive processes also becomes increasingly crucial. This allows us to critically evaluate these technologies, considering their potential benefits and limitations.

CHAPTER NINE
Buying Blind Spots

Shoppers in a busy Portuguese supermarket stared in astonishment at the middle-aged woman pushing her heavily loaded trolley around the aisles. What attracted their puzzled attention wasn't what the woman was buying but what she was wearing. A skull cap bristling with electrodes fitted snugly over her long, dark hair to record her brain activity as she browsed the shelves. A pair of eye-tracking glasses identified what brands and products she attended to, those she ignored and the order in which she viewed them.*

To see what she saw, go to lewisandleyser.com/eye-tracking or scan the QR code.

She was taking part in the world's first-ever study to analyse what happens in a shopper's brain when he or she scans supermarket shelves: where that person looks, what they see and how they respond to varying brands. Similar studies were later conducted in Berlin, Boston, Dublin, London, New York and Paris. Although the shoppers' nationalities were

---

* Here it is worth noting that gaze (the direction we look in) and fixations (the time spent focused on a specific object) don't always tell us with 100 per cent accuracy what we are paying attention to. Individuals may look directly at an object but still fail to perceive it because their attention is elsewhere.

different, their shopping behaviour largely remained the same: while shoppers *looked at* around 50 per cent of the products, they *saw* only some 20 per cent while paying close attention to less than 5 per cent.[1]

Research suggests there are two main reasons why blind spots made eight out of ten products invisible to shoppers. First, the items were, at least at that moment, of no personal interest or relevance to them. And second, the shoppers were covertly encouraged to ensure their limited attentional resources were not 'squandered' on anything of lower financial benefit to the retailer.

Despite extensive research and substantial marketing budgets, 75 per cent of the more than 1 million new products launched each year fail. A third of the 120 new products launched in the top five European markets and the 90 in the US each day will have disappeared from the shelves within a few weeks, at a cost of some £6 trillion annually.[2]

While some products, such as Crystal Pepsi, Jimmy Dean Chocolate Chip Pancake-Wrapped Sausage, Clairol's Touch Of Yogurt Shampoo, and Watermelon Oreos,[3] seemed destined to fail from the start, the majority do so not because they are disliked, distrusted or deliberately ignored, but because little or no attention is paid to them.

We spend our lives in what John Browning and Spencer Reiss, editors of *New Economy Watch*, describe as an 'attention economy' – a glittering haze of commercial messages in which the quantity of information is effectively infinite, but our access to it is limited by both the time and mental energy available. Forty years ago, supermarkets stocked around 7,000 items. Today, they invite shoppers to choose between 40,000 and 60,000; Amazon offers readers 2 million titles. Including both films and TV programmes, the Netflix library has well over 17,000 titles globally and over 5,000 available at any one time.[4]

Retailers and advertisers have long believed that greater choice equates to higher sales and more satisfied customers.

Over the past twenty years, however, research has cast increasing doubt on this seemingly common-sense conclusion. 'As the number of choices keeps growing, negative aspects of having a multitude of options begin to appear,' writes psychologist Barry Schwartz in *The Paradox of Choice*. 'At this point, choice no longer liberates but debilitates.'[5] And it's the same with every type of would-be persuasive message, from charitable appeals to political campaigns. If we register these at all, it is only fleetingly and in such a way that little of the expensively generated information remains in our memory.

In 2000, Sheena Iyengar and Mark Lepper conducted a series of studies showing that although consumers initially seemed intrigued and empowered by more choices, this enthusiasm quickly faded. In one study, they set up a tasting booth for exotic jams at Draeger's, a gourmet grocery store in California. As customers passed the tasting booth, they encountered either a limited display of six different jams or an extensive display with 24. The number of passers-by who approached the tasting booth and the number of purchases made in these two conditions were carefully recorded.

Although extensive choice proved initially more enticing than limited choice, limited choice was ultimately more motivating. Thus, although 60 per cent of passers-by approached the table when there were 24 varieties of jam on offer, compared to only 40 per cent when there were only six, 30 per cent of those offered the limited selection purchased a jam, while only 3 per cent of those offered the extensive selection did so.[6]

In the same study, Iyengar asked participants to select a chocolate from among either six or 30 others. Before making their choice, subjects were asked to predict whether their chocolate would taste 'excellent' or merely 'satisfactory'. After tasting their selection, they provided actual ratings of satisfaction. It was found that those invited to choose from six different chocolates were far more satisfied and likely to purchase than participants who chose from 30 different options. 'Collectively, these results suggest that choosers may experience frustration with complex choice-making processes and that dissatisfaction with their choices may lead to a lower willingness to commit to one choice,' comments Iyengar.

Since that study, over a hundred others involving more than seven thousand participants have confirmed the negative impact of choice overload. While consumers initially regard an abundance of choice as highly appealing, it reduces sales and increases dissatisfaction with purchases.

For consumers, then, one of the most effective ways of overcoming blind spots is to deliberately reduce one's search to no more than six options. For retailers, meanwhile, the aim should be to create a positive emotional response between the customer and the brand. 'The customer won't understand a bad experience consciously,' says consumer psychologist Kate Nightingale. 'They will have this "off" feeling, this slight discomfort, and it will make them leave your store or abandon your website. The more instances of these "off" feelings they have, well, that can get them to leave your brand for good.'[7]

Grocery shopping, for many consumers our most regular form of consumption, is where we run the greatest chances of having blind spots manipulated. Apart from the vast choice, shopping in a supermarket differs from other types of consumerism in two ways: stress and boredom. 'Grocery shopping is a chore most shoppers dread,' says retail marketing expert Emily Rodgers. 'Overcrowded aisles, long checkout lines, and difficulty finding certain items can make the thought of visiting a grocery store unbearable.'[8]

The level of stress consumers experience while supermarket shopping depends on various factors, an important one being the amount of choice on offer. The more choice there is, and the less time people have to choose, the greater the vulnerability to inattention-, choice-, change- and expectation-related blind spots.

Boredom, meanwhile, comes into play due to the frequency and repetitive nature of the task – and this too can cause blind spots. A recent study involving 2,000 adults found that they visit retail stores three times a week for over 37 minutes each time – and that doesn't include the additional 22 minutes it takes to get to and from the supermarket.[9]

As a result, many shoppers trade speed for accuracy.[10] They make purchasing *choices* rather than *decisions*. Although often used synonymously, 'deciding' and 'choosing' differ in several important ways. Decisions, primarily the product of rational thinking, take between minutes and months to make. Choices, typically made in as little as three thousandths of a second, are powerfully influenced by emotions and more vulnerable to attentional misdirection. The main difference between the two is the amount of attention each requires. People making tough decisions show greater dips in attention reserves than those making choices.

The attentional energy required for choosing or deciding varies according to age, gender, health, familiarity with the task, mental state (i.e., tired or rested, sober or drunk) and level of emotional arousal. The more a person's reserves drop, the greater their risk of blind spots. Attention reserves are rapidly depleted when many decisions must be made in a limited time. Decision fatigue sets in, and the risk of experiencing blind spots increases. People often 'go with their gut' if no decision has been made as their energy reserves run low. They plump for a quick choice, sometimes with disastrous consequences. When based on intuition, choices produce faulty results due to unrecognised biases, leading to various blind spots. Manipulating someone's blind spots to encourage a switch from slow and rational thinking to fast emotional thinking helps market billions of products, services and political ideas each year.

While making a TV documentary on supermarket shopping, we interviewed people as they left a shop to see whether they had indulged in impulse purchases. All but one of the twenty questioned admitted having done so, with spur-of-the-moment purchases including picture frames, potted plants, a teddy bear and a cactus. The only customer in our study who hadn't made an impulse buy turned out to be a supermarket manager, who had made a written list of items to purchase and stuck to it because, as he put it, he 'knew the tricks of the trade'. To watch this interview, go to lewisandleyser.com/shopping or scan the QR code.

Store knowledge is another reason blind spots flourish among supermarket grocery shoppers. 'Store knowledge determines the extent to which product and brand search is guided by internal or external memory,' says professor of marketing Choong Whan Park. 'When consumers shop in a familiar store, search is guided primarily by internal memory that requires minimal effort and thus facilitates the performance of in store decision-making activities … when consumers have little knowledge of a store's layout, search activities must be guided by external memory (e.g. in-store information displays) that requires considerable effort. This, in turn, reduces a consumer's ability to perform other in-store decision-making activities.'[11]

The less familiar a shopper is with a store's layout and the position of products in aisles and on shelves, the greater the demands on their working memory and the more rapidly their attentional reserves will become exhausted. Rather than searching their memory for information relating to a particular proposition, they use the first relevant piece of information that comes to mind, whether or not it is perceived as accurate.

Studies have found that people remember implausible arguments better than plausible ones and are more likely to recall these as a basis for judgement. Thus, more accessible information may be used as a basis for judgement even though it is not necessarily more believable.[12]

The tedium and stress of shopping – not to mention the cost and hassle of transportation – are primarily responsible for the rapid rise in online buying and selling. In 2022, customers spent $992 billion on purchases through social media, with social media commerce expected to reach $8.5 trillion by 2030.[13]

This does not mean, of course, that selling online is without its problems – far from it. Research analyst Cara Malone from Juniper Research has estimated that the global cost of online fraud, where cybercriminals conduct false or illegal transactions using strategies such as phishing, business email compromise, or account takeover, could reach $206 billion cumulatively between 2021 and 2025, with businesses standing to lose more than $343 billion.[14]

Nor is fraud the only challenge facing online retailers. Nine out of ten customers abandon their shopping trolleys if the checkout is too complicated. Walking away from their purchases is slightly more likely among older than younger consumers (90 per cent of those aged 55+ compared with 83 per cent of millennials), with the latter also being more likely to return to the same website later (12 per cent of millennials compared with 7 per cent of those 55+). When shopping in store, seven out of ten UK customers abandon even fully laden shopping trolleys if they wait six minutes or more at the checkout. American shoppers are somewhat more patient, waiting up to eight minutes before walking away, with 77 per cent less likely to return to a store where they have experienced long checkout delays.[15]

This underlines the importance of the acronym KISS (Keep It Straightforward and Simple) for retailers, advertisers and marketers. The more attention-demanding and time-consuming a sales message, the greater the stress and the more likely a product will remain unseen.

Inside the Selling Machine

Supermarkets are not so much shops where people buy things as sophisticated machines designed to sell them. An environment in which – as with a magician's tricks – consumers are most likely to see only what the retailer wants them to see while remaining largely blind to everything else.

'The typical customer is assumed to walk up and down the aisles of the store, stopping at various category locations, deliberating, choosing the best option, and then continuing in a similar manner until the path is complete,' says Jeffrey Larson.[16] Market researcher Herb Sorensen disproved this widely held view by attaching radio frequency identification (RFID) tags to supermarket carts. He found that the average shopper only visits around 25 per cent of the shop and typically follows one of fourteen routes around the aisles.[17]

Most shoppers begin by moving around the store's outer perimeter in what Sorensen termed 'the racetrack'. There is less congestion on the outer edges of the aisles, and they can move faster. Rather than exploring the entire length of each aisle, they make short excursions in and out of only a few. Because of this, products displayed in the middle of aisles tend to receive less attention than those placed nearer each end.

The shelf location where products stand the greatest chance of selling varies according to whether they are what retailers term 'utilitarian': necessities such as food and clothes, or 'hedonic': products associated with fun, pleasure, and excitement, satisfying our wants rather than meeting our needs.

The third and fourth shelves up are the most favourable height for utilitarian items, associated with an 8.3 per cent and 10.6 per cent sales increase, and the worst for 'hedonic' products, associated with a 22 per cent decrease in sales. These self-pampering products actually do best on the bottom shelf.

Figure 77

Products at 1.25 metres and higher are least likely to be seen, and any on the seventh shelf, around 1.50 metres from the floor, are 18 per cent less likely to sell.[18]

The image above shows how one of the shoppers in our research, 36-year-old mother-of-three Mary, scanned the supermarket shelves for breakfast cereals. Larger circles indicate where the most attention was paid to a product, with a greater likelihood of those items becoming part of what retail specialists Judith and Marcel Corstjens call the consumer's 'Mindspace'.[19]

The Power of Persuasion

Why specific products will be sought out and given our attention while others remain unseen is a question that has engaged the brightest minds in advertising, retailing and marketing for decades. While there is no simple answer, one aspect that research and practical experience have shown to be of great importance is priming: the techniques used to ensure that consumers are subtly encouraged to look only where the retailer wants them to look and see only what they want them to see.

In September 1957, a cinema audience in New Jersey watched a film in which messages priming them to 'drink Coca-Cola' and 'eat popcorn' flashed briefly on to the screen at a subliminal level of perception – that which, while discriminated by the senses, fails to reach conscious awareness and cannot be verbally reported. Although those watching the film were never consciously aware of the phrases popping up, market researcher James Vicary said these messages increased popcorn sales in the foyer by 60 per cent and Coca-Cola sales by 20 per cent.[20]

News of this experiment, conducted at the height of the Cold War, led to a horrified reaction from press and public alike. One journalist described it as 'the most alarming and outrageous discovery since Mr Gatling invented his gun'.[21] A second nonsensically advocated the discovery be 'attached to the centre of the next nuclear explosive scheduled for testing'.[22]

Five years later, in an interview with *Advertising Age*, Vicary admitted his 'study' had been a hoax designed to boost his failing marketing business. For most psychologists and advertisers, his deception appeared to hammer the final nail into the coffin of subliminal priming for advertising purposes.[23] One writer described it as an 'urban myth'.[24]

However, the fact that the phenomenon as originally presented by Vicary and Vance Packard in their bestselling book *Hidden Persuaders* did not stand up to scientific scrutiny was by no means the end of the story.

The idea of subliminal priming is often misinterpreted by the general public, with the phrase being incorrectly applied to advertising techniques involving images of scantily clad people, vulgar words and dominant trade colours. Since these elements are consciously attended to, no subliminal priming is involved.[25]

Research going back over a century, however, has found that actual subliminal priming not only works but exerts a powerful sway over what shoppers see and buy, influencing our purchasing decisions in a host of subtle but profound ways. Advanced data analytics, neuroscience and psychology-based practices enable marketers to craft highly personalised and emotionally powerful campaigns that tap into consumers' subconscious desires and motivations.

In a recent study, psychologists presented participants with nine slides of people engaged in a routine daily activity. These were preceded by slides showing positive or negative events, such as a small child happily playing with a doll versus a shark ripping into a dead seal. None of the emotionally charged slides were consciously perceived since these were only shown for a few thousandths of a second. When later asked to evaluate the target person, it was found that those preceded by a 'happy' image were evaluated more positively and rated as having a more agreeable personality than those preceded by a sad one.[26]

In another study, participants were shown a series of geometric shapes for a thousandth of a second — far too brief an exposure for them to be consciously aware of. They were then presented with pairs of shapes, one shown previously and another never seen before. Researchers found that, while unable to say which of the two they had seen, they showed a greater liking for the shape subliminally presented to them.[27]

Results from studies like these have led to multi-million-dollar product placement deals on major movies and popular TV shows. Marketing and advertising executives pay between £46,000 and £195,000 (US$60,000-250,000) for actors to use or wear a product such as a car, a camera, a computer, a fashion item or a soft drink.

One of film history's first major product placements was the candy used to lure a shy little alien out of the woods in Steven Spielberg's 1982 film *E.T. the Extra-Terrestrial*. After M&M turned down the producer's original placement offer,

Hershey used the slot to promote Reese's Pieces, their new peanut butter candy. While the exact amount of money that the movie placement made for them is unknown, insiders estimate it generated a 300 per cent increase in sales.

Similarly, in the 1999 film *The Matrix*, Neo, played by actor Keanu Reeves, used a Nokia 7110 mobile phone to escape arrest. 'After *The Matrix* became a huge hit,' says marketing guru Brock Diedrick, 'so did Nokia's 7110 phones.'[28]

Although usually no direct reference is made to the product, and most viewers probably never notice it, the subliminal familiarity provided by product placement increases the chances of the product being purchased. 'In the light of all the evidence,' says psychologist Ap Dijksterhuis, 'we may safely conclude that attitudes can be changed (or formed) by subliminal evaluative conditioning.'[29]

Land on Mars – Buy a Mars Bar

When, on 4 July 1997, NASA's Pathfinder spacecraft landed on Mars, it captured worldwide media attention and stayed in the headlines for months. During the same period, sales of the chocolate bar Mars, named after the company's founder, rather than the planet, dramatically increased.[30] 'This,' says Professor Jonah Berger, 'was a lucky turn of events for the candy bar company, to be sure, but what does it mean for our understanding of consumer choice?'[31]

Essentially, it reveals the extent to which aspects in our surroundings that we are unaware of profoundly influence what we look at but fail to see.

In a study by Berger, people were randomly handed an orange or a green pen to answer beverage-based questions before being offered a free drink of either Fanta or Sprite. He found that, without ever being aware of the way the ink colour was biasing their choices, they were more likely to select an orange-coloured can of Fanta if they had used an orange pen and a green-coloured Sprite can if they had used a green one.³²

Another study looking at the importance of environment was conducted by behavioural psychologist Travis J. Carter and his colleagues who, after controlling for voting patterns, positioned a small and easily overlooked Stars and Stripes at the upper edge of a computer screen where volunteers read information unrelated to politics. The researchers found that the flag influenced voting patterns, with participants more likely to vote Republican than Democrat up to eight months later.³³

This tallies with findings that national flags are a pervasive priming cue in the political landscapes of many countries. They can appear on houses, schools, and government buildings and, in the US, on the lapels of politicians and those seeking office. Flags constitute incredibly potent political cues because, without necessarily being consciously noticed, they may still reinforce nationalist sentiments. Although there may be occasions when a national flag unifies people, there are many instances in which it nudges them towards one or the other end of the political spectrum. Under these circumstances, the flag may bias the citizenry towards a particular political party or product, often without their awareness.

Priming With Words

Semantic priming, a phrase first used in the early 1970s, refers to the consistent finding that people respond faster to target words preceded by related, relative to unrelated, primes.³⁴ For example, we recognise the word DOUGHNUT faster when preceded by COFFEE than by FRENCH FRIES. Likewise the word NURSE if it follows DOCTOR rather than BREAD.³⁵

In terms of advertising, repeated exposure to a company's name or logo, a familiar package design or a political slogan can influence consumers without their awareness. Triggered memories can significantly change people's behaviour, as a study by psychologists Esther Papies and Petra Hamstra demonstrated.

To test the potency of semantic priming, they posted a low-calorie meal recipe, claimed as being 'good for a slim figure' on a butcher's shop door. Once inside, customers were offered a tray of beef meatballs and invited to eat all they wanted. Even though a minority of customers consciously noticed the recipe, it still influenced how much the majority ate. On the days it was displayed, fewer snacks were eaten than when it was absent. The unobtrusive little dieting poster had 'overruled ... the tendency to overeat'.[36]

In another study, by neuroscientist Bahador Bahrani, subjects wore glasses that projected faint pictures of everyday objects (such as pliers and a domestic iron) to one eye and a strong flashing image to the other – a technique known as 'continuous flash suppression', which works to erase subjects' awareness of the pictures seen. At the same time, subjects performed either an easy task – picking out the letter T from a stream of letters – or a task that required more concentration: picking out a white N or blue Z from the same stream. While images did register during the easier task, during the more demanding task, an fMRI scan failed to detect any image-related neural activity.[37]

This finding – that the brain does not pick up on subliminal stimuli if it is too occupied with other things – shows that some degree of attention is needed for even the subconscious to pick up on subliminal images. 'Your brain does log things that you aren't even aware of and can't ever become aware of,' Dr Bahrani reports. 'There is a brain response in the primary visual cortex to subliminal images that attract our attention – without us having the impression of having seen anything.'

Familiarity and Exposure

During the winter of 1967, a mysterious student wearing a black fabric bag from the top of his head to his bare ankles began attending psychology lectures at Oregon State University. According to lecturer Charles Goetzinger, 'Bag Man' was an ordinary student who wanted to remain anonymous.

Initially resented as a distraction, Bag Man was subjected to some hostility. One student poked him with an umbrella and tried to pin a 'Kick Me' label on the bag, though as the weeks passed, most of his classmates accepted and became protective of him, shielding him from prying journalists. After one term, he disappeared and, to this day, his identity remains unknown.[38]

Half a century later, this story lives on in psychology and marketing lore as an example of the 'mere exposure effect' popularised in 1968 by Robert Zajonc. Zajonc cited the mysterious Bag Man as illustrating that people will grow to like a thing they initially detest – even with no new information – if they are repeatedly exposed to it.[39]

As the old saying goes: 'We may not always know what we like, but we tend to like what we know'.

Research has shown that even when exposure to a message is too fleeting to catch someone's full attention, it is sufficient to increase their preference. 'This effect is most likely to occur when there is no preexisting negative attitude toward the stimulus object, and it tends to be strongest when the person is not consciously aware of the stimulus presentations,' report marketing specialists Tom Stafford and Anthony Grimes. Even if an advertisement, logo or brand is noticed only briefly, it still benefits from being noticed at all.[40]

We are continually exposed to brands in our daily lives, not only as a result of marketing activities but also as a consequence of encounters with others. Some of these exposures are long-lasting and involve direct communication and engagement, as when we notice the brands and discuss them with others. Others occur briefly and in passing. Shoppers glimpse another person's brand selections at the supermarket checkout line. Drivers admire a new car as it passes them. We see people walking by and wearing the latest sports shoes.

Information gleaned from these brief exposures has been shown to substantially impact the observer's later choice of that brand. 'Because the brand is generally not the focal point of the encounter, the exposure to the brand itself is incidental in nature, and any processing of brand information in these encounters is likely to be nonconscious,' says Rosellina Ferraro, an assistant professor of marketing.[41] Ferraro and her colleagues refer to these as incidental consumer brand encounters or ICBEs. They believe they can exert a potent, yet little appreciated, effect on which products we see. This was demonstrated in an experiment using different bottled water brands. One was Dasani, a well-known brand launched by Coca-Cola.

While familiar to study participants, Dasani was neither the most popular nor dominant brand on campus. Participants were shown photographs of fellow students in various everyday situations, such as waiting in a bus shelter, walking to lectures, or lying on the grass while studying. In each case, the participant was instructed to focus on facial expressions, to minimise the likelihood of them noticing that a bottle of Dasani water was included in some pictures but absent from others.

As a 'thank you' for participating, they were offered a complimentary bottle of water and invited to choose from four brands: Dasani, Aquafina, Deer Park and Poland Spring. The more often they had been exposed to the Dasani brand, the more likely they were to choose it. In the zero-exposure control condition, Dasani was selected by 17.1 per cent of the participants, yet for those exposed to it four times, preference rose to 21.6 per cent, and for those with 12 exposures preference rose to 40 per cent.

Information unconsciously gleaned from such brief exposures helps 'activate its representation in memory and generate fluency', explains Rosellina Ferraro, leading people to regard it more positively and choose it more frequently. Because these ICBEs are commonplace occurrences, the ease with which people can process the information has been proposed as the mechanism underlying the effect.

The Unseen Power of Shrinkflation

We have already examined how on-shelf placement can cause inattentional blind spots, impacting the chances of a product catching the consumer's eye or remaining unseen.

However, change blind spots, sometimes combined with illusional blind spots, can also impact the retail environment. One example is the profit-boosting tactic American economist Dr Philippa Malmgren has called 'shrinkflation'.[42] This is when manufacturers, primarily in the food and beverage industries, boost profits by reducing the size or quantity of products.

According to the UK's Office for National Statistics (ONS), between the beginning of 2012 and June 2017, 2,529 products shrank in size. Packs of Maltesers, M&Ms, and Minstrels were reduced in weight in the UK by 15 per cent. Walkers Crisps similarly removed two bags from its 24-pack but kept the price the same. In the US PepsiCo replaced their 32oz (950ml) bottle with a 28oz (825ml) one while increasing costs. A roll of kitchen paper, which previously held 600 sheets, now offers only 320.[43]

Of course, shrinkflation could go badly if spotted, turning consumers off the product and encouraging them to change brands. But in practice, change blindness means most will remain unseen most of the time. Many grocery shoppers operate in a fast, less attention-demanding thinking mode. They are popping the item straight into their basket or trolley rather than pausing to read and reflect on labels disclosing the truth.

Illusional blindness can also play a role. In one instance involving washing-up liquid, manufacturers introduced a kink into the middle of their bottles, arguing it made it easier to hold and pour. However, this change also enabled them to reduce the amount of liquid while camouflaging the shrinkage by having the bottle remain at the same height.

Manufacturers of the world's best-known and most popular brands spend vast sums building a product's reputation. They strive to ensure it will be instantly identifiable through unique packaging, colours, logos and taglines. This enables copycat products to reap the benefits of established and recognised brands while avoiding reputation-building expenses.[44] And thanks to change and expectation blind spots, many consumers eagerly purchase these lower-cost items, often without noticing any differences and expecting them to equal the quality of the originals.

While copycat brands undoubtedly seek to exploit consumers' preferences for the familiar, this does not mean shoppers are always unaware of the mimicry. Although many consumers won't notice the brand is a knock-off, others will have made a conscious buying decision for budget reasons. Copycats are not only cheaper and often not that different from the branded product, but in some instances similar branding aids recognition of the product type. Having a similarly shaped bottle and coloured packaging for a bottle of off-brand Irish cream liqueur, for instance, will help shoppers identify it on the shelf as a liqueur rather than a bottle of wine.

CHAPTER TEN
Business Blind Spots

In 1975, the board of the century-old Eastman Kodak company was offered an opportunity that could have made its fortune; instead, it contributed to its bankruptcy. Understanding how and why this happened is fundamental to appreciating the potentially devastating role of blind spots in business.

Electrical engineer Steve Sasson joined Kodak at 24, weeks after graduating with honours from the prestigious Rensselaer Polytechnic Institute. Keen to find him a task that would, in his own words, 'keep me from getting into trouble doing something else',[1] Kodak instructed him to investigate charge-coupled devices (CCD). Invented by Bell Labs researchers Willard Boyle and George E. Smith in the late 1960s, these converted light into electrical signals. Steve's bosses wanted to know whether they had any practical use in photography.

After months of tinkering, Sasson discovered that they most certainly did. He found a way of digitising the electronic pulses they produced to record, store and view full-colour images. 'This was more than just a camera,' explained Steve. 'It was a photographic system to demonstrate the idea of an all-electronic camera that didn't use film and didn't use paper, and no consumables at all in the capturing and display of still photographic images.'

Despite the prototype's clunky appearance and poor-quality pictures that could only be viewed on a television set, Sasson was confident Kodak directors would recognise the camera's money-making potential and be eager to invest in its development. To the young inventor's dismay, instead of seeing his camera as an opportunity, they considered it a threat to their long-established and hugely profitable film and printing paper business. Rather than exploring and exploiting its potential, they focused only on the device's technical problems: the 23 seconds required to save each image on to a cassette tape, the poor picture quality and the fact that you could only view them on a television. With modestly priced colour prints readily available, his bosses could not see why anyone would want to look at photos on their TV.

Following his brief presentation to Kodak's board, Sasson was ordered never to publicly discuss his invention or demonstrate the prototype to anyone outside Kodak. His camera was locked in a cupboard, where it remained for the next 14 years.

'The main objections came from the marketing and business sides,' says business journalist John Estrin. 'Kodak ... made money on every step of the photographic process. If you wanted to photograph your child's birthday party, you would likely use a Kodak Instamatic, Kodak film and Kodak flash cubes. You would have it processed at the corner drugstore or mail the film to Kodak and get back prints made with Kodak chemistry on Kodak paper.'[2]

So by the time Kodak produced their first digital camera in 1989, it was too late. An ever-increasing number of photographers, amateur and professional, were abandoning film for the instant gratification of digital photography and had already turned to Kodak's competitors. In 2012, the 125-year-old company filed for bankruptcy with liabilities of $6.75 billion.[3]

As with Kodak, everyone managing or working in a rapidly changing business world will face numerous opportunities and endless problems. How successfully they recognise and pursue those opportunities and solve those problems depends on the strategies they adopt and the mistakes they avoid. Dr Graham Edkins estimates humans make between three and six errors per hour regardless of the activity or task.[4] While many are inconsequential, others prove more serious and costly, as shown by these three marketing mistakes.

In 2009, Tropicana changed their standard packaging from an image of a fresh-looking orange to favour a more minimalist image of a glass of juice. This simple rebrand was quickly recalled, but only after sales fell by 20 per cent, costing the company millions of dollars.

If Tropicana had tested the design's associations and brand fit beforehand, they may have reconsidered its use. When deciding to change the design, they should have done so incrementally to not break the law of just noticeable differences, which in logo design refers to the extent to which products can change their logos while maintaining brand recognition.

A second marketing mistake was demonstrated by Google in 2013, when they released their prototype Google Glass, a wearable, hands-free smartphone that displayed information in the wearer's field of vision and could be controlled by voice command. They had impressive marketing and an amazing idea that looked like the future, but the product flopped when a lack of familiarity and concerns over privacy made many uncomfortable about buying the product.

Our third example comes from March 1985, when Coca-Cola released their first formula change in 99 years. 'New Coke' was promoted as replacing tried-and-tested regular Coke. It was meant to re-energise the Coca-Cola brand, whose share lead over Pepsi had been slipping for 15 years. However, consumers were outraged by the change because of the generations of continuity between Coke and its classic flavour. After a public firestorm in which Coca-Cola received over 400,000 letters and calls of complaint, they returned to the original recipe in July 1985, just 79 days after the change. This was a case of loss aversion – a cognitive bias whereby people perceive a situation as worse when framed as a loss rather than a gain. In this case, the loss was a taste they had become familiar with. Although New Coke was better liked in taste tests, the benefits needed to be stronger to outweigh the loss of the original product.[5]

Confronting the Challenge of Change

How people approach the challenge of change and how willing they are to take risks when seeking solutions to novel problems depends to a great extent on their personalities and learning experiences. On whether they approach the unknown with the mindset of what the 6th-century BCE Greek soldier and poet Archilochus described as a 'fox' or a 'hedgehog'.[6]

'Foxes seem to know something about everything while hedgehogs seem to know everything about something in particular – that one big trick,' says professor of strategy and intelligence studies, Randy Borum. 'Throughout history, there have been some phenomenal hedgehogs whose single, unifying big ideas transformed entire fields of enquiry. It was true for Freud and the concept of the unconscious; for Marx with his idea of class struggle; and with Darwin and his proposed process of natural selection.'[7]

Political science writer Professor Philip Tetlock sees the fox/hedgehog concept as a way of understanding the differences between two ways of thinking, managing and leading.[8] Great leaders and scientists are often hedgehogs. They have one big idea, shunning or disregarding complexity and seeking straightforward answers to even the most challenging problems. Winston Churchill was a hedgehog who never let contradictory information interfere with his set ideas. This led him to be right about Hitler. He was never distracted by the contingencies that might combine to make the elimination of Hitler unnecessary. The upside of being a hedgehog is that you can be really and spectacularly right. The downside is that you can also be catastrophically wrong.

In a stable and unchanging business world, doing what you've always done on the assumption you'll get similar results can prove a viable management model. However, in today's rapidly changing business world, expectation blind spots can quickly become a recipe for financial and commercial disaster.

The Kodak leadership appear to have adopted a hedgehog approach to changing business trends. This led the directors to believe photography's future would be the same as photography's past. When they decided to hide Sasson's invention in 1987, film had been used, processed and printed in much the same way for over a hundred years. No one was complaining, and prints were inexpensive. Given this background, it is easy to see how change and expectation blind spots should dominate their thinking. 'We like to solve problems easily,' points out psychologist Gordon Allport. 'We can do so best if we can fit them rapidly into a satisfactory category and use this category to prejudge the solution ... It takes less energy and effort.'[9]

On the other hand, foxes, says Philip Tetlock, 'see the world as a shifting mixture of self-fulfilling and self-negating prophecies: self-fulfilling ones in which success breeds success, and failure, failure but only up to a point, and then self-negating prophecies kick in as people recognize that things have gone too far.'[10]

Such an approach might seem favoured when running a company, since the risks can be so great and the stakes so high. Yet blind spots are still present in a task every manager and many employees must undertake, usually many times a day: problem-solving. There are two ways in which these blind spots arise.

The first is by causing people to remain blind to the fact that there is any problem to solve. This is especially likely when the problem lies outside the problem-solver's previous experience or comfort zone, circumstances that make it more likely to be dismissed as unimportant or irrelevant, or seen as a threat rather than an opportunity.[11] 'Managers frequently pretend not to see the problems,' says business consultant Jerry Osteryoung 'Not only does this affect the morale of the entire office, it undermines the manager's authority. The rest of the staff will eventually lose respect for the leader because failing to take action against bad behaviour is tantamount to accepting it.'[11]

The second way blind spots arise in decision-making is through a failure to understand that problem-solving is both an ability – requiring natural talent, training and experience – and a process requiring a series of steps. More than fifty years ago, psychologist Wayne Wickelgren discovered that every problem, no matter how complex, has three essential components: *givens*, *operations* and *goals*.[12]

Failing to realise the components of problems is a frequent cause of wrong answers and confused thinking. Before trying to find the best solution to any business problem, you must answer these three questions:

Do I understand the *givens*?
Do I understand the *operations*?
Do I understand the *goal(s)*?

Because these three components are so important, let's consider each in more detail.

Givens

Givens comprise everything we know or can find out about the problem. Failure to do this adequately is a frequent cause of mistakes and misjudgements. The problem solver may ignore, overlook or fail to discover an essential given because it is assumed from general knowledge rather than expressly stated.

Research has shown that ignoring or dismissing givens is most frequent in strongly hierarchical companies or organisations, where information provided by anyone lower down in the 'pecking order', especially if that person is a woman, will be dismissed or ignored. Professor of communication Kathleen Propp demonstrated the extent of this gender-induced blind spot in a 1995 study. She assembled a group of men and women ostensibly to make decisions in a child custody case, a subject in which women might be expected to have greater knowledge than men.[13] While the entire group was given information about the family concerned, a few were provided with a piece of additional information — an important 'given' the remainder did not have. When introduced into the discussion by a man, it was six times more likely to be seen as relevant and acted upon than when introduced by a woman.

A tragic example of the bias against information provided by women occurred in Israeli intelligence in the weeks leading up to Hamas's massacre of civilians on 7 October 2023. According to the liberal Israeli newspaper *Haaretz*, three months before the attack, Mai, a female intelligence officer working for Unit 2800 of the Israeli Defence Force, warned her commanding officer about two platoons of Hamas fighters conducting a military exercise. An expert regarding the terrorist organisation, she reported: 'They were practising crossing the border into Israel, raiding a kibbutz, attacking a military academy, and killing all the cadets.' Hamas had ended the drill with the words, 'We have completed the murder of all the residents of the kibbutz.' Far from taking her warning seriously, superior officers dismissed her detailed report, describing it as 'imaginary'.

'They abandoned our friends to die,' said one of her female colleagues sadly. 'Nobody wanted to listen to us. It's beneath their dignity. Who am I, some little woman, before a man with the rank of major or lieutenant colonel, for whom everybody stands at attention when he enters the room? Nobody really pays any attention to us.'[14]

When considering the givens, then, it is important to consider suggestions and information from everyone qualified to provide them. Never dismiss ideas, facts or figures from someone due to their age, gender, ethnicity or position in the company.

It is also important to avoid inattentional blind spots caused by 'functional fixedness'. This term, coined by the psychologist Karl Duncker during the 1940s, describes how people will often only use an object in the way it's traditionally used rather than thinking outside of the box and using items in novel ways. In one experiment, Duncker gave his subjects a candle, a box of matches and a few drawing pins. Their task was to find a way of attaching the candle to a wooden door so the melting wax did not drop to the floor. Only a minority succeeded.[15] Can you think how they could have done it? See the end of the chapter for a successful technique.

A good way of uncovering every significant given is by brainstorming, a problem-solving strategy devised by American advertisement company manager Alex Osborn in 1938.[16] Brainstorming sessions should be informal, non-hierarchical and spontaneous to stimulate creativity and encourage diverse opinions. Everybody should adopt a 'yes and' rather than a 'yes but' approach to every given, judging ideas not solely on their current status but their potential. Sessions should be time-limited, with everyone present encouraged to develop ideas and suggestions. No matter how unlikely they seem, all should be noted.[17]

Only once you are sure of your givens and have ensured you have not fallen victim to choice, change or inattentional blind spots can you be sure of embarking on your operations in a way most likely to lead to a successful outcome.

Operations

Operations are the processes by which givens are manipulated to find the best solution or achieve a desired outcome. A major stumbling block to success is a form of tunnel vision that sets the mind in an intellectual rut and causes it to try and solve the problem by familiar but, on occasion, inappropriate tactics.

Imagine, for example, you were given the problem of using six matches to create four equal-sided triangles with no other shapes allowed. You might like to try this before reading further. If you haven't got six matches to hand, try visualising the operations you might perform on them – operations that must not include snapping any of them in half! You'll see the correct operation at the end of this chapter, together with an explanation of why people so often fail to find it.

Before you arrive at a solution, ensure that your blind spots have not prevented you from identifying all possible ways the givens might be manipulated. Only once you are satisfied can you turn to the problem's final component, your goal.

Goal

In *convergent* problems, where there is a single correct answer, such as 2 × 2 = 4, the goal is precisely defined. Most problems in business, however, are *divergent*. They have many possible answers, although usually only one or two best ones. With such problems, it may prove impossible to devise satisfactory operations for achieving one sort of goal but possible after viewing that goal in another way. If, for example, your goal to *raise* an object has proved impossible, the same result might be achieved by *lowering* something else.

The value of reframing a goal is illustrated in this classic problem from Arabia.

A wealthy merchant had two sons who prided themselves on their horsemanship and the speed of their stallions. Each boasted that his mount was superior to the other, and his skill in the saddle outshone his brother's abilities. At length, the merchant, growing tired of their endless arguments, offered a wager to settle matters once and for all. His entire fortune would be bequeathed to whichever of the young men won a race from his palace to the city, some hundred miles away across the desert. But, instead of the prize going to the man whose horse reached the city gates first, it would go to the son whose mount arrived *last!*

The riders set off, each travelling as slowly as possible so that after several days they were still only a few hundred yards from their father's home, and it seemed likely that the race would last forever. Indeed, such might have been the case but for the arrival of a wise man who, after being told the nature of the challenge, was able to give the men some advice. Seconds later, they galloped as fast as they could towards the city.

What words of wisdom did the old man offer?

In this problem, the answer lies in neither the givens (two horses and riders) nor the operation (to ride as slowly as possible) but in the goal. Reflect on how you might quickly and easily ensure one of the horses reached the city last without the riders having to go as slowly as possible. As before, you'll find the answer at this chapter's end.

To demonstrate how correctly understanding givens, operations and goals is essential for successful problem-solving, consider the following: a candle is 15cm long, and its shadow is 45cm longer. How many times longer than the candle is the shadow?

If you said three times longer, your answer corresponds to that given by 87 per cent of those we asked. Yet you are mistaken. The given states that the shadow is 45cm longer than the candle and, therefore, four times as long.

Congratulations if you got the correct answer, but if – like the vast majority – you made a mistake, do not feel bad about it. Once you understand the three components of that and every other problem, you will be unlikely to make such a mistake again.

By listing the givens, operations and goals sequentially, we may have led you to believe they always operate in that order. Viewing them as forming an endless circle would be far more accurate. Goals often determine the givens selected and the operations employed. Operations can also determine the goals achieved and the givens chosen.

To put all this to the test, try discovering the secret behind this mind-reading trick known as the Amazon prediction.[18] It appears to require paranormal powers until you apply your knowledge of its givens, operations and goals to work out the straightforward way it is done. Here's what happens.

A magician begins by quickly scrolling through Amazon on his phone, emphasising that over 350 million products are available. Placing the phone on the table, he asks you to select an item from the online store and write it on a blank card without revealing your choice. Let's say you chose a blue Chinese vase.

The magician then asks a few seemingly unconnected questions, such as your star sign, favourite colour and whether you prefer cats or dogs, explaining that your answers will help him identify the chosen product.

As you reveal your choice, the magician turns his phone to show you the screen. On it are a picture and description of the blue Chinese vase you selected. You can see a video of this trick performed by Keelan, by going to lewisandleyser.com/prediction or by scanning the QR code.

But how did he know which one of the 350 million items would be chosen?

This problem has three givens: the smartphone, card and pen. Since you cannot inspect them, please take our word that all were exactly as they seemed and not 'tricked up' in any way.

Regarding possible operations, it may help to watch the video to see if you can spot the very simple manoeuvre that allows the magician to 'read' what was written on the card. Take inattentional blind spots into account by asking yourself what the magician did to distract you while he searched Amazon for the chosen item.

The goal is obvious: to discover how the magician, short of being a psychic, could have known which item would be chosen.

If you want to check your theory or see how the trick is done, go to lewisandleyser.com/prediction-explanation or scan the QR code.

To return to the Eastman Kodak story, their given was Sasson's digital camera. The operations they should have performed include developing, manufacturing, marketing and selling the revolutionary new product. In the event, the company only patented the camera before hiding it away for decades, instead of pioneering digital photography and riding the wave of prosperity it would deliver. Their goal remained to exploit their monopoly in manufacturing and marketing films, printing paper and processing, confident these would remain hugely popular for at least another hundred years – an example of expectation blind spots in action.

In fairness to the Kodak board, few could have foreseen the speed and thoroughness with which digital technology would transform photography and every aspect of life.

Problem-Solving in the Digital Age

As the Second World War ended, IBM boss Thomas Watson predicted the world would never need more than eight computers![19] He could not have been more wrong. Today, there are over 2 billion computers, including servers, desktops, and laptops and 9 billion smartphones and tablets worldwide.[20]

Not only are there now more computerised devices than people on Earth, but they are also becoming smaller, faster and more 'intelligent' with every year that passes.

The Frontier supercomputer at Oak Ridge National Laboratory in Tennessee, for example, can perform 2 quintillion calculations (2×10^{18}, or 2 followed by 18 zeros) in less than a second. (This is twice the level of performance defined as an *exaflop*.) Using its vast computing power – the equivalent of 100,000 laptops – scientists can instantly come up with solutions to problems that would take the world's entire population four years to solve.[21] And despite this, Oak Ridge engineers are designing a new supercomputer three to five times faster than Frontier, to be unveiled in the coming decade. An auto industry journalist has worked out that if cars had developed at the same pace as computers, they would have 66,764,192 horsepower, go from 0 to 60mph in 0.0034 seconds, get 3,666,652 miles per gallon, and cost only US$4,471![22] As a result, the quantity of new knowledge, both reliable and fictitious, is increasing exponentially. It has been estimated that more than 2.5 quintillion (2.5×10^{18}) bytes of data are generated daily, with this figure increasing as a result of ever-growing internet use.[23]

The more information the brain has to process, the greater the risk that blind spots will cause crucial facts to be unseen or misunderstood. In July 2023, some 75,000 passengers were stranded abroad for days when 1,500 flights were cancelled after two identically named but separate location markers were erroneously entered into the air traffic control computer. Faced with the dilemma of either approving or rejecting the flight plans, the system crashed.[24]

The difference between mistakes made by computers and humans is that when silicon brains blunder, steps are swiftly taken to ensure they never happen again. When humans do the same, they are more likely to seek refuge in denial ('it didn't happen on my watch') and identify scapegoats ('it was someone else's fault').

In a study of how people respond to mistakes made by computers compared to those made by fellow human beings, Andrew Prahl and Lyn M. van Swol asked participants to perform an unfamiliar task: scheduling hospital operations. To assist them, they were given help from either an 'advanced computer system' or someone experienced in operating-room management. Sometimes the advice dispensed was sound, but at other times it was very wide of the mark.[25]

The researchers found that participants 'punished' the computer after receiving lousy advice by refusing to use it again. When their human advisor blundered, they were more likely to move on from the mistake and then trust the person as much as before. 'It is almost as if people "forgave" the human ... for making a mistake,' says IT specialist Andrew Prah, 'but did not extend the same feelings of forgiveness to the computer.'

This finding has important implications for the future of work, where computers are increasingly replacing humans. As the researchers point out: 'Any potential efficiency gains by moving towards automation might be offset because all the automation has to do is err once, and people will rapidly lose trust and stop using it.'

Although technological glitches and computer gremlins occur, human error is often responsible for business blunders when people or corporations fail to see what is there or when they see what is not there. To paraphrase Shakespeare, in our increasingly sophisticated high-tech world, 'the fault, dear Brutus, lies not in our machines but in ourselves'.

In today's ever more digitised workplace world, business blind spot errors almost always come down not to errors made by computers themselves, but to a conflict between our carbon-based brains and the computers' silicon-based ways of 'thinking'. Professor of law and philosophy Scott Shapiro has termed this the conflict between 'upcodes' and 'downcodes'.[26]

Upcodes are generated by everything above our fingertips — from the inner operations of the human brain to the outer social, political and institutional forces that define the world around us. They shape our thoughts, mould our beliefs, determine our behaviours and include social and cultural codes, personal morality, religious rituals, social norms, legal rules, corporate policies, professional training, our experiences, and ethics that influence us from within and, often invisibly, from without.

The term downcodes refers to everything beneath our fingertips when typing on a computer keyboard – the technical codes that enable the device to function. These include the microcode embedded in chips, device drivers for your printer, operating systems such as Windows, Linux, and iOS, and high-level programming languages such as C and Java.

There are several different techniques for preventing upcode errors from leading to downcode failures. The first line of defence is avoidance: this is typically accomplished through user interface design or training. In the former case, the user interface is constructed to block potential errors. For example, programs can guide users through predefined tasks, or human input can be removed entirely via automation. Neither approach tends to be very successful in practice since operators often bypass both when trying to accomplish tasks unanticipated by the user interface's designer.

Instead of blocking errors at the interface, an alternative is to train human users not to make errors. Training works by developing the human mental model of the computer system, thereby preventing mismatches that constitute a significant source of error. It is only as effective as it is extensive, however; training must focus on concepts – not just procedures – to help build the broadest mental models, and it must evolve with the system to ensure that those models are kept up to date. The best training programs are extensive, frequent and designed to force operators out of their comfort zones; technology can help achieve these goals both through regular training and by developing software robust enough to ward off cyber-attacks.

One such program, developed by Netflix, is called Chaos Monkey. This program disrupts the critical functions of the software. The challenge facing software engineers is to devise a system capable of resisting and recovering from the operating changes caused by such attacks.[27]

In today's rapidly evolving business landscape, emerging technologies like Artificial Intelligence (AI) and big data analytics are reshaping how companies make decisions. While these tools offer unprecedented benefits, from enhanced efficiency to predictive insights, they also introduce new cognitive blind spots that business leaders must navigate. Just as Kodak failed to see the digital revolution coming, many companies today risk overlooking the subtle yet profound ways AI is transforming their industries – and potentially their own decision-making processes.

Let's explore some specific examples of how these emerging technologies are reshaping the business landscape and potentially introducing new cognitive blind spots.

Blindspots and Computers

One way in which this emerging technology is potentially introducing new cognitive blind spots is in businesses becoming overly dependent on AI predictions, ignoring human intuition and expertise. This was seen in the 2008 financial crisis when overreliance on scientific risk assessment models contributed to the market crash.

People tend to trust automated systems more than human judgement, potentially leading to uncritical acceptance of AI-generated recommendations.

AI systems trained on historical data may perpetuate existing biases. For example, Amazon's AI recruitment tool showed bias against women because it was trained on predominantly male resumes.[28]

Another issue comes from data-quality blind spots, whereby data analytics can lead to false conclusions if the data is incomplete or of poor quality. For instance, Google Flu Trends initially overestimated flu prevalence due to limitations in its data sources.[29]

The use of big data can also create blind spots around customer privacy. Facebook's Cambridge Analytica scandal is a prime example of how data misuse can lead to major ethical and business issues.[30]

Other blind spots may include overlooking new cybersecurity risks, such as adversarial attacks on AI systems; 'black box' decision-making, in which AI systems make decisions that are difficult for humans to interpret or explain, potentially creating regulatory and ethical blind spots;[31] and data silos, whereby different departments in large organisations may use separate AI systems, leading to inconsistent decision-making and missed insights from data integration.[32] AI systems optimised for short-term metrics may also overlook long-term consequences, creating strategic blind spots.

Integrating AI and big data into business processes, then, is not just a technological shift but a cognitive one. It requires leaders to develop new mental models and decision-making frameworks. As we've explored, these technologies can both illuminate existing blind spots and create new ones. The key for businesses moving forward is to harness the power of AI and big data while maintaining a critical awareness of their limitations and potential biases. This balanced approach — combining technological capabilities with human insight and ethical considerations — will be critical in navigating the complex business landscape of the future. By staying vigilant to these new forms of blind spots, companies can position themselves at the forefront of innovation and sustainable success.

To illustrate how the various problems described above can combine with intentionally engineered blind spots to steal billions of dollars from investors, consider one of the 21st century's most outrageous financial scandals – the rise and fall of billionaire Samuel Benjamin Bankman-Fried, founder of the cryptocurrency exchange FTX.

Born on 5 March 1992, American entrepreneur Bankman-Fried became crypto's poster boy and one of the wealthiest people on Earth. Before his catastrophic fall from grace in 2022, SBF, as he was known, had been ranked as the 41st-richest American by *Forbes*. However, in December of that year, he was arrested in the Bahamas and extradited to the United States. There, the US government brought civil and criminal charges against him and some of his top executives for misappropriating over US$8 billion in customer deposits, laying the groundwork for charges of wire fraud, commodities fraud, securities fraud, money laundering and violation of campaign finance law. Convicted, he will now spend the next 25 years in prison. His acquisition of so much wealth, admiration and respect owes much to his ability to utilise magic strategies in high finance. So which strategies helped him the most?

Anyone attempting to manipulate free will rather than to compel compliance by means of terror, as occurs in autocratic regimes, must start by either persuading people to like them, or represent themselves as the kind of person others would like to be.

Known as the 'similarity-attraction' effect, this was first described by psychologist Theodore Newcomb in the 1960s. He reported that college roommates who shared similar attitudes and values were likelier to develop close friendships than those who did not.[33] Since then, numerous studies have confirmed the existence of the similarity-attraction effect across various social contexts and populations, including friendships, romantic relationships and work environments.

Salespeople, politicians, magicians and confidence tricksters widely employ the strategy of making it seem that they like those they're trying to influence. This does not necessarily mean being personally known, liked and trusted, however; achieving celebrity status or receiving celebrity endorsements will do equally well. This was one of the strategies used by SBF.

A charismatic and personable young man, he cultivated close ties across the political spectrum, using charm and money to gain influence. Celebrities and public figures like Bill Clinton were paid handsomely to boost his image and enhance his credibility. Clinton was reportedly paid over $250,000 to speak alongside SBF at a crypto conference in the Bahamas. The Clintons also provided him with a platform at their annual Clinton Global Initiative conference, endorsing him just months before FTX's collapse.[34] SBF donated nearly $40 million to United States Democratic candidates and causes leading up to the 2022 midterms. However, the public at the time was unaware he had also donated vastly more untraceable 'dark money' to Republicans.

According to Damian Williams, the US Attorney for the Southern District of New York, 'All of this dirty money was used in service of Bankman-Fried's desire to buy bipartisan influence and impact the direction of public policy in Washington.' While these contributions were made to look as if they had come from wealthy co-conspirators, they were funded with stolen customer money.[35]

Just as pseudoscience tricks people into rationalising magic, SBF's political donations tricked customers into rationalising investment despite red flags about his rise and claims. While it is undoubtedly true that just because celebrities allow themselves to be bought does not necessarily mean they like or admire the person with the cash (only that they like the money!), this makes no difference to the outcome. As SBF had anticipated, the halo effect brought him a high liking and trust among potential investors.

SBF used lawyers and accountants to create complex, jargon-laden arguments to provide a plausible explanation for where the missing customer money went. In magic, a 'false solution paradigm' exploits people's tendency to rely on initial explanations without reflecting on why that explanation was presented. SBF hijacked this mental process as well.

We are living in what some palaeontologists have called the Anthropogene, or 'Age of Man'. While all organisms influence their environments to some extent, few have changed the world as far, or as fast, as our species is doing. Global warming, pollution, the destruction of habitats are among the reasons for the ongoing mass extinction of plant and animal species with experts predicting that, if carbon emissions continue to rise unchecked, up to half of plant and animal species in the world's most naturally rich areas, such as the Amazon and the Galapagos, could face local extinction.[36]

Yet the Anthropogene has also seen a vast increase in human knowledge and digital technology, now being used to understand these changes, predict their effects, and prevent or ameliorate the damage.

Whether or not they are able to do so will, to a great extent, depend on which problems we seriously attend to and which go unseen and unattended.

In the next chapter we will be exploring why some of these problems may go unseen.

BUSINESS BLIND SPOTS

Answers to problems

Affixing a candle to the door: After emptying out the matches, use one of the drawing pins to affix the now-empty matchbox to the door and use it to hold the candle. Duncker reports that when the matches were placed separately on the table, so that the matchbox was seen as only a container, far more people could solve the problem.

Figure 78

Making four triangles using six matches: This problem can only be solved by breaking out of the blind spot that forces you to reason in two dimensions. Instead of moving the matches around on a flat surface, build a pyramid, with three forming the base and three on the sides. This gives you a total of four equilateral triangles.

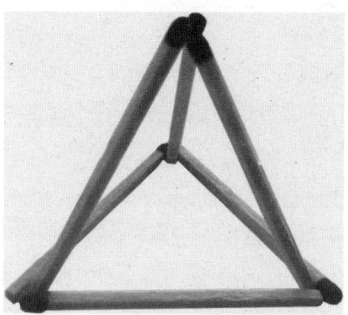

Figure 79

The wealthy merchant's horse race: Since the winner would be the one whose *horse* was last to arrive at the city gates, the old man advised the two riders to swap horses!

CHAPTER ELEVEN
Magic Blind Spots

Keelan balances a dessert spoon between his thumb and forefinger. As his audiences watch in disbelief, the spoon's stainless steel handle slowly folds in half without his applying any pressure, just as it did when first performed during the 1970s by ground-breaking illusionist Uri Geller.[1] As with Norman Triplett's 'vanishing baseball', explanations for how this happens range from the unlikely to the impossible, with some of those watching insisting he must possess psychic powers!

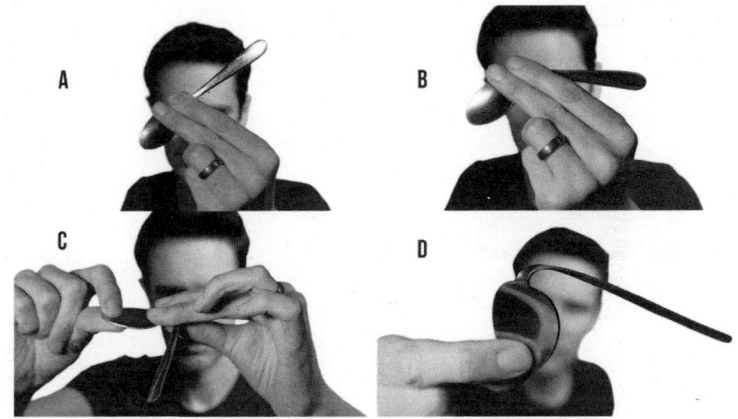

Figure 80

Watch Keelan perform the spoon bend at lewisandleyser.com/spoon-bend or scan the QR code on the left overleaf. To see one way in which the trick can be done, go to lewisandleyser.com/spoon-bend-explanation or scan the QR code on the right.

'The method by which a trick is performed should be invisible,' says magician Barrie Richardson. The fewer moves or gimmicks needed, the more elegant and magical the effect will be.[2]

No longer solely the preserve of magicians, virtually undetectable visual manipulations are used for a wide range of commercial, social and political purposes – for example, to sell products, promote brands, advocate social policy or entice voters. As William Freudenburg and Margarita Alario point out: 'The politician's versions of disappearing acts are best understood not by asking about what it is we see ... but by learning more about what it is we fail to see, and why certain techniques are so effective at keeping problematic questions unasked, forgotten, or hidden in the shadows, away from public view.'[3]

Choices We Are 'Forced' to Make

'Pick a card,' the magician invites. 'Any card. The choice is yours.'

A spectator does so, taking great care to ensure no one else sees it. Moments later, when the magician correctly names that card, the audience wonders how anybody could have known which of the 52 cards had been selected.

The answer is simple.

The spectator's 'free choice' was nothing of the sort. The magician named the card because that was the card the spectator had been manipulated into choosing. You can view one of the numerous 'forcing' techniques used by magicians at lewisandleyser.com/riffle or by scanning the QR code on the next page.

MAGIC BLIND SPOTS

'Although we like to think that we are using our free will to make our decisions,' says psychologist Alice Pailhès, 'we are often oblivious to the cognitive mechanisms that underpin our decision. Many of our behaviours are automatic and unconsciously influenced by external stimuli.'[4]

Magicians call the technique of covertly controlling another's free will a 'force'. It is the art of determining someone's choice without them being aware of the extent to which they were compelled to make it. 'Indeed,' says former magician Jay Olson, 'the magical effect typically evaporates once the spectator realises their decision has been influenced. A lack of awareness of the force is essential.'[5]

Let's examine the most powerful and widely used ways to employ a force: *distractions*, *misinformation*, and *priming*.

Distractions

On a busy summer beach, Keelan and his partner Richard Reed invite holidaymakers to participate in a 'love test'. This involves seeing how high they can bounce a rubber ball. The higher it bounces, they claim, the more attractive that person is to the opposite sex.

But while the ball bounces sky high for both the magicians, much to the chagrin of the young men and women taking part, it refuses to bounce for them. Focused on the ball, participants remain oblivious as Keelan 'steals' the watch off their wrists, which you can watch by going to lewisandleyser.com/pickpocketing or scanning the QR code on the next page.

When carrying out their 'thefts', the magicians employ one of magic's most ancient and powerful techniques – distraction.

Hieronymus Bosch's painting *The Conjurer*, from the early 1500s, illustrates this method. While a magician distracts an audience's attention with a cup and ball trick, his young accomplice steals an onlooker's money. The trick is as relevant today as it was 500 years ago.

'Many stage illusions involve attentional misdirection, guiding the observer's gaze to a salient object or event, while another critical action, such as sleight of hand, is taking place,' explains psychological researcher Anthony

Figure 81

Barnhart. 'Even if the critical action takes place in full view, people typically fail to see it due to inattentional blindness.'[6]

Indeed, it has been found that we can only simultaneously follow a maximum of four moving objects while holding the same number in working memory.[7] 'As a result,' notes neuropsychologist Georgia Gregoriou, 'we are only aware of a limited number of objects, typically those that are a subject of our attention.'[8]

And distractions aren't only used by pickpockets and illusionists. Roman emperors were known to create elaborate 'circuses' to divert the attention of the population from their unpopular social and political issues.[9]

Today, replicating the principles of the ancient Roman circus on a grand scale involves all forms of music, sport, entertainment and especially social media. The world now spends in excess of 720 billion minutes per day using social platforms. This amounts to more than 260 trillion minutes, or 500 million years, of collective human time each year.

'The "typical" internet user spends almost 2.5 hours each day using social media platforms, equating to more than one-third of our total online time,' says digital marketing expert Simon Kemp. 'On average, that means that social media accounts for 35.8 percent of our daily online activities, meaning that more than *1 in 3* internet minutes can be attributed to social media platforms.'[10]

Much of this time is spent on what is referred to as 'doom-scrolling': the compulsive desire to continuously scroll through bad news or negative content on social media and websites.

'Tech companies are masters of manipulation', point out marketing academics Kaitlin Woolley and Marissa Sherif. 'Endless feeds, autoplay videos, and those red notification dots are all carefully crafted to keep us glued to our screens. Algorithms keep you on your phone for longer by serving you content they predict you'll like through your engagement habits. If the content feels relevant and interesting to you, not only do you continue to scroll to find more of it, but you're also more likely to open your apps through the anticipation of finding something you enjoy.'[11]

While private companies own and profit from these modern-day versions of the Roman circus, governments determine the political environments in which they operate and are by no means averse to using them to distract from bad political news or unpopular policies.

Increasing numbers of people use social media as their primary, even sole, means of entertainment, amusement and information.

Research by Jianxun Chu and colleagues have found this is associated with a fear of missing out (FOMO) on rewarding experiences enjoyed by others. FOMO can rapidly create an unhealthy attachment to the phone, resulting in an inability to resist the urge to reach for it during idle moments – a dependency fuelled by the instant gratification of social media, games and other apps. Seven in ten people admitted they would feel depressed, panicked or helpless if their mobile device was lost or stolen.[12]

As if it had a closed-circuit TV in every room of every house on every city street, the internet sees everything you do online, including when you do it and how long you are doing it for. Whenever you load a page in your browser, what you see (and much of what you don't) comes not only from that address but from different elements from different places. Whether you like it or not, every time you accept 'cookies' from a website, a small file of data is sent to your device, which tracks and stores your information. Algorithms are then able to use that data to determine what you later see.

With the adverts and other material suggested for you constantly tweaked in response to fresh data, these are designed to attract as many clicks as possible and keep the user online for as long as possible. Although the person who clicks on these external links could technically be argued to be looking for information, they have, in reality, been attracted by something unlikely to be giving them any helpful information.

The extent of social media usage depends significantly on age, with a pronounced gulf in understanding between what technology writer Marc Prensky terms 'digital natives' and 'digital immigrants'. The former, born after the start of the digital era, a date set in 1980, will have spent their entire lives surrounded by and using computers, video games, digital music players, video cameras, smartphones and all the other toys and tools of the digital age.[13]

Many commentators believe that among this group social media increases political participation, boosts political interest, generates and sustains political knowledge, and re-activates citizens who are not politically involved.[14]

Digital immigrants were born into an analogue world where computers, if they existed at all, were large, costly and cumbersome machines only understood by highly trained specialists. Given this they are often confused by and sometimes fearful of a technology they are unable to fully, or even partially, understand.

In their study of political and entertainment-oriented content on social media, professor of communication science Jörg Matthes and his colleagues discovered differences between participants based on the amount of time devoted to entertainment versus that spent seeking reliable information. Although focused on political information, we believe their findings to have a broader application.[15]

In the lower left quadrant, those termed Hermits score low on accessing both information and entertainment. Described as 'the inactive' by Matthes and his colleagues, this group tends to be older, less well-educated and lacking support from computer-savvy younger relatives.[16] That said, although social

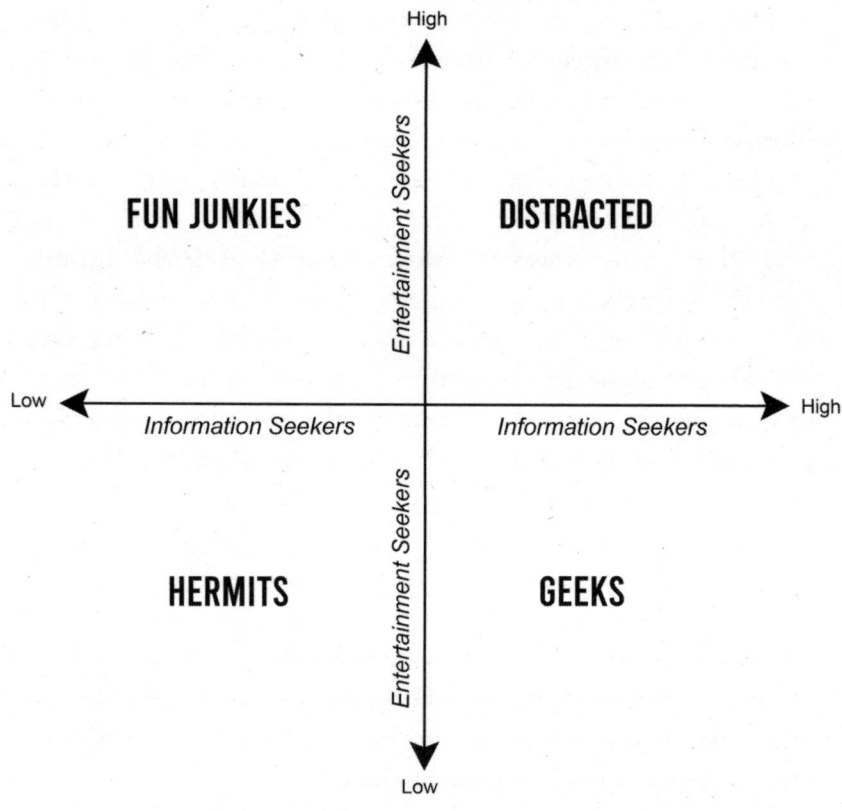

Figure 82

media plays little or no part in their lives, they may obtain political and other information through family, friends and work colleagues who themselves do get their information from social media.[17]

Studies indicate there are many more Hermits than one might think. In this generation four out of ten US adults never use Facebook; 81 per cent do not use Instagram, while 92 per cent don't use X (named Twitter at the time of the study) and 81 per cent don't use WhatsApp.[18]

With the wide range of media available, it is easily possible for people who care little about political debates to focus solely on entertainment. This group, termed Fun Junkies, situated in the top left quadrant, actively avoid the news, question its credibility, feel overwhelmed by its sheer volume, or dislike its predominantly negative focus. Watching TV, scrolling social media, playing video games and messaging friends form an integral part of their lives online, with as little time as possible spent seeking information.

'When an individual exposes him- or herself to a limited amount of news because other content has more appeal to him or her,' say researchers Morten Skovsgaard and Kim Andersen, 'algorithms will make future content decisions in favour of other content, e.g., entertainment, and give news stories a lower priority.'[19] This results in their becoming locked into a pattern in which their enthusiasm for being entertained leads to being offered more opportunities to satisfy that enthusiasm.

In the upper right quadrant are the Distracted users. Although motivated to seek entertainment and information to an equal extent, they often allow the former to distract them from paying sufficient attention to the latter. Attempting to simultaneously check social media, play games, surf the web, listen to music, watch TV and chat with friends, the Distracted are confident they can switch from one activity to another without losing accuracy. Studies have shown such confidence to be seriously misplaced. 'It takes time (an average of 15 minutes) to re-orient to a primary task after a distraction such as an email,' reports cognitive psychologist Paul Atchley. 'Efficiency can drop by as much as 40 per cent. Long-term memory suffers, and creativity – a skill associated with keeping multiple, less common associations in mind – is reduced.'[20]

Researchers have also discovered a significant relationship between multitasking and changes in the brain. Neuroscience researchers Kep Kee Loh and Ryota Kanal used functional magnetic resonance imaging (fMRI) to study the brains of 75 adults and found heavier media multitaskers possess smaller amounts of grey matter in their anterior cingulate cortex

(ACC), the brain region involved when paying attention, making decisions, anticipating rewards, controlling impulses and detecting errors. 'Changes to their brains reduced their ability to focus on relevant information while inhibiting task-irrelevant stimuli,' explains Dr Loh.[21]

Finally, in the lower right quadrant are those almost entirely motivated by a desire to seek reliable political information. They have little interest in entertainment. 'Such people', say Skovsgaard and Andersen are 'rarely exposed to entertainment-oriented forms of social network.' Because their newsfeeds are primarily information-based, they are unlikely to be distracted by entertainment-oriented forms of social networking.[22] These individuals are termed Geeks.

Although Jörg Matthes's research focused on social media and politics, our experience suggests Fun Junkies, the distracted, hermits and geeks exist among consumers and audiences at magic shows, with Fun Junkies and the distracted being most susceptible to distractions of all kinds.

Forces based on distraction are widely used by business and political interests to persuade customers and voters only to see what they want them to see. You will encounter a commercial distraction each time you visit a restaurant. After placing your order, a waiter will arrive with freshly sliced bread whose true purpose is to distract diners from the time taken to serve their meal. The head waiter at an exclusive restaurant in London's West End told us: 'As soon as they have some bread to occupy themselves, diners stop bugging staff who are busy taking care of other tables. It stops them from becoming a right *pain in the ass*.'[23]

Misinformation

A second widely used magic technique is misinformation, in the form of false evidence appearing real (FEAR). Hollowed-out coins, double-sided playing cards, stacked decks, mirrored production boxes, hidden magnets and invisible threads produce mistaken assumptions that enable the magician to subvert reality by manipulating the audience's blind spots.

Away from the world of entertainment, those motivated by a desire to manipulate the choices of others resort to even more complex ways of creating an illusion of false reality – their 'props' range from deepfake images to unsubstantiated tweets and WhatsApp messages. Computer science researcher Soroush Vosoughi and his colleagues report that for all forms of news, false news propagates faster and more widely than real news, with the problem being particularly evident for political news. The most pernicious fake news stories were routinely read by 10,000 people, while true stories were rarely read by more than 1,000.[24]

During the run-up to the Taiwan elections, in January 2024, users of YouTube, X, Instagram and other social media sites were bombarded with videos entitled 'The Secret History of Tsai Ing-wen', making false claims about the outgoing president and her ruling party. Fronted by Chinese- and English-speaking news anchors, these were shared 100 times a minute at their peak and widely believed by young Taiwanese, some of the world's most technologically sophisticated people. All the videos were false, probably posted by state-backed Chinese propaganda organisations known as Storm-1376, Dragonbridge or Spamouflage. As for the smooth-talking anchors, they were AI creations made using CapCut, an app from TikTok's parent company Byte-Dance.

And this is a global problem. Cognitive psychologist Sam Wineburg and his team tested the ability of thousands of UK and US students to distinguish between fake and genuine information using the Misinformation Susceptibility Test (MIST),[25] the product of four years' work by Rakoen Maertens and Friedrich Götz.

Participants were asked to decide whether 20 statements, such as: 'Left-Wing Extremism Causes "More Damage" to World Than Terrorism, Says UN Report' and 'Hyatt Will

Remove Small Bottles from Hotel Bathrooms', were real or fake. Their answers indicated their tendency towards scepticism or gullibility.[26]

To the surprise and concern of Wineburg and his colleagues, over 80 per cent of high school students failed to distinguish truth from falsehood. For example, 70 per cent considered an article sponsored by an oil company that branded global warming a hoax more reliable than a peer-reviewed scientific paper. Contrary to the widely held belief that older, less digitally savvy 'digital immigrants' were the most likely to be taken in by fake news, they found digital natives were far worse. Among 18- to 29-year-olds, only 11 per cent could distinguish between real and fake news, compared to 36 per cent of those aged 65 or older.

'Susceptibility to misinformation depends on much more than just factual knowledge or education,' reports Rakoen Maertens. 'Lots of people share fake news, not necessarily because they actually believe [it], but rather because they belong to a social group, or they have a group identity or attitudes towards a certain issue, and they want to express that attitude.'[27]

The more time either digital natives or digital immigrants spent on social media, the less likely they were to distinguish genuine news from misinformation. Over 53 per cent of those who got news from Snapchat failed the MIST test, with just 4 per cent getting high scores. Users of Truth Social, WhatsApp, TikTok and Instagram were just a little behind when it came to their susceptibility to fake news and misinformation (45 per cent, 4 per cent, 41 per cent and 38 per cent, respectively).

'The problem is much bigger than just fake news,' says social psychologist Sander van der Linden. 'We are increasingly dealing with viral half-truths, deeply partisan agendas, and constant media manipulation … this type of fake news only represents a tiny fraction of misleading media more generally.'[28]

Take the story of £860,000 (US$1.1 million) worth of jewellery allegedly purchased by Ukrainian First Lady Olena Zelenska. On 30 September 2023, a young woman posted a video on YouTube claiming to have been working as an intern at Cartier's New York store when the president's wife allegedly splurged money from Western aid funds on the jewels. A receipt was shown briefly, and the story ended by saying the sales girl had been fired.

A few days later, the story, wholly untrue and easily disproved, was published on numerous West African and Russian news sites and discussed on Western talk shows. Posted to X, it was reposted and shared more than 20,000 times. And yet Olena Zelenska was actually in Canada that day, not New York, and the 'Cartier intern' was, most likely, a St Petersburg beautician. Despite this, the tale quickly went viral and was presented as further 'evidence' of Ukraine's misappropriation of money intended for the purchase of weapons.

Discussing the findings of his studies, Vosoughi says that: 'Tweets containing false information were more novel – they contained new information that a Twitter user hadn't seen before – than those containing true information. They elicited different emotional reactions, with people expressing greater surprise and disgust. That novelty and emotional charge seem to be what's generating more retweets.'[29]

And correcting misinformation and 'fake news' through verification websites has been shown to make matters worse rather than better, as social data scientist Mohsen Mosleh and his colleagues discovered when they corrected fake news stories on X (then Twitter), such as the allegation a US District Court judge had ruled that girls in an Illinois school district had no right to privacy and must 'shower with boys'. Rather than ensuring the misinformation vanished from social media forever, the corrections increased the retweets and made the language used even more partisan.[30]

The global spread of misinformation has exerted a significant negative impact on society. Conspiracy theories about vaccines have been linked to a growing reluctance to use them, with an inevitable rise in preventable diseases such as measles and TB. Misinformation about the impact of 5G has led to the vandalisation of cell phone masts, while climate change misinformation has been associated with a reduction in perceptions of scientific consensus.[31]

Raised alongside computers, digital natives, the oldest of whom will, at the time of writing, now be in their mid-forties, see the world in a distinctively different way from anyone who grew up before the information age got going. 'They develop hypertext minds,' says Marc Prensky. 'They leap around. It's as though their cognitive structures were parallel, not sequential.'[32]

However, while such cognitive skills may prove invaluable in the digital world, they make digital natives far more susceptible to blind spot manipulation through misinformation.

Misinformation not only affects how we see current events but also our memories of how we saw events in the past. The more time that has passed since an event, the more vulnerable memories of what happened will become, and the greater the chances it will be contaminated or even replaced by misinformation. While such memory manipulation can occur at any age, its effects are more pronounced among the young, with children and teenagers proving especially susceptible. In a study demonstrating the extent of this susceptibility, psychologist Professor Elizabeth Loftus used misinformation to implant a false memory in the mind of Chris, a 14-year-old boy.

She told him that, at the age of five, he had wandered away from his mother and older brother in a busy shopping mall. Lost for over an hour, he was in tears when an elderly man reunited him with his family. Despite the account being fictitious, Chris claimed to clearly 'remember' the toy store where he got lost, that the bald-headed man who rescued him was wearing an old blue flannel shirt, and how his mother had scolded him for running off.[33]

In another study, psychologist James Ost and his colleagues implanted a false memory in a middle-aged woman. They informed her how, at the age of four, she had been diagnosed as having low blood sugar. Although at first doubtful, she became increasingly convinced the incident had occurred and developed an ever more vivid 'memory' of this non-existent event: 'It must have been on a Sunday because my dad was there,' she told researchers. 'He was always around on a Sunday ... My mum was panicking quite a lot, and I kept passing out ... my dad must have rushed me to the hospital ... it is a massive, huge place. They did a blood test on me and found out that I had low blood sugar.'[34]

As intriguing, if ethically questionable, as these examples of the fallibility of individual memories are, of greater interest – especially to anyone whose business centres on creating memorable images for others – is the Mandela effect. This phenomenon demonstrates that no matter how precise our recollection of something may be and how confident we are about what we have seen, there is an excellent chance of it being wrong.

The term 'Mandela effect' was coined in 2010 by paranormal researcher Fiona Broome after she noticed that many, herself included, falsely remembered Nelson Mandela dying in prison in the 1980s when he passed away in 2013.[35] While accepting this as an interesting phenomenon, psychologists rejected her explanation of alternative timelines crossing in parallel universes in favour of these being examples of false memories and social contagion: the extent to which we unconsciously mimic the emotions of those around us. Smiles, for example, can quickly spread from one person to another, as can expressions of sadness or disgust.

Psychologists Deepasri Prasad and Wilma Bainbridge conducted a fascinating study into the visual Mandela effect (VME), in which one image is mistaken for the correct version, using cultural icons such as the Monopoly man and Pikachu, Pokémon's mascot for over 25 years.[36] They created two altered images of 40 brand logos and cartoon characters; for example, versions of the Monopoly man augmented with either a monocle or spectacles. Shown the three images – one original and two modified – participants were asked to choose the correct image and say how confident they were in their choice. Five of the 40 cultural icons, including the *Star Wars* character C-3PO, the Fruit of the Loom logo, the Monopoly man and the Volkswagen logo, consistently fell victim to the visual Mandela effect, with an overwhelming majority selecting the same incorrect image and expressing a high degree of confidence in their choice.

In another experiment, Prasad and Bainbridge used eye-tracking to establish that there were no significant differences in looking patterns between the items the participants got correct and incorrect. This demonstrated the VME was due to neither a lack of attention nor unfamiliarity with the original, leaving expectation blind spots as the most likely explanation. We incorporate expectations about how things *should* look into our memories of items; for instance, a monocle is incorporated into our visual memory of the Monopoly man since such wealthy, elderly men often wore a monocle as a sign of their upper-class membership.

This ties into how our preexisting mental frameworks can create blind spots, causing us to 'see' or remember what we expect rather than what's there. The fact participants were often unable to explain why they made VME errors also highlights how cognitive blind spots usually operate below our conscious awareness.

This study also raises intriguing questions about the nature of memory and perception, challenging our assumptions about the reliability of our recollections. This connects to the broader concept of overconfidence bias, a common cognitive blind spot where we overestimate the accuracy of our knowledge or memories. The fact that the VME occurs in both recognition and recall tasks suggests that these blind spots can affect multiple cognitive processes and may be culturally specific, so that cultural context plays a part in shaping our cognitive blind spots. While the study didn't find evidence that attention differences cause VME, the phenomenon still demonstrates how our memories diverge from reality despite seemingly clear perceptual experiences.

Awareness of the Mandela effect might draw attention to specific false memories, making them more salient in people's minds; however, this increased focus could inadvertently reinforce these misconceptions, making it harder for individuals to distinguish between accurate memories and shared false ones. Simply being aware of our potential for false memories doesn't necessarily protect us from them and might even make us more susceptible in some cases.

While VME usually arises spontaneously, deepfake photographs have also been used to implant false memories. When shown a doctored picture of Bugs Bunny in a Disney World advertisement, one adult in six 'remembered' meeting and shaking hands with the rival Warner Brothers character.[37] Studies like this have shown the extent to which, once implanted, misinformation can be quickly elaborated into a detailed false memory believed to be so accurate it triggers people's emotions and actions.

Interested in understanding what happens at the neurological level during the creation of false memories, neuroscientists Yoko Okado and Craig E. L. Stark scanned the brains of 20 students as they watched short videos, each comprising 50 images. Twelve of these vignettes contained an item that changed across time and served as misinformation of the events.[38] When asked two days later which images they remembered, participants were able to recall the altered images more often than the original versions, suggesting the creation of false memories. By examining their brain images, the researchers observed that the interaction of encoding processes in two regions at the front of the brain, the medial temporal lobe and the prefrontal cortex, are critical in determining whether true or false memories are reported.[39]

And it's not only in a laboratory setting where false memories occur – there are real-world examples of people developing false memories for implausible or impossible experiences, such as being abducted by aliens.[40]

When memories are retrieved and reconstructed, distortions may creep in without explicit external misinformation. Neurologist Heike Schmolck and her colleagues tested this by using one of the 1990s' most notorious murder trials to see how well people's recollections stood the test of time.[41]

In June 1994, 47-year-old football player and film actor O. J. Simpson was charged with the murders of his ex-wife Nicole Brown and her friend Ron Goldman. Acquitted after a lengthy and highly publicised trial, he was found guilty of their killings in a civil case three years later.

The Simpson case was, for months, the leading topic on television news and the print media. Three days after the verdict, Schmolck asked college students about the trial, repeating the same questions and noting the answers 15 and 32 months later. After 15 months, 50 per cent of the recollections were highly accurate, with only 11 per cent containing significant errors or distortions. By 32 months, however, accuracy had fallen to 29 per cent, with more than 40 per cent containing major distortions. These results underline the significant changes in recollections occurring over time.

These studies reveal how, by implanting vivid and detailed false memories, misinformation can come to be regarded as more accurate and reliable than the truth.

'The obvious question,' says Elizabeth Loftus, 'is why we would have evolved to have a memory system that is so malleable in its absorption of misinformation?'[42] The answer, she believes, lies in the fact that the 'updating' of memories due to misinformation is the same as the perfectly regular and valuable updating that occurs when new information causes us to modify or correct our recollections. This is essential if we are to operate in a rapidly changing world. The downside is it also provides a 'back door' allowing outsiders to manipulate how people remember the past and see the present. It is a vulnerability that makes implanting false memories into the minds of millions simple and more concerning than ever.

Priming
While distractions and misinformation are potent ways of manipulating blind spots, priming, or 'perception without awareness' as we prefer to call it, can be an equally powerful, if more controversial, way of controlling what people see or fail to see.

Psychologist Boris Sidis was among the first to demonstrate the power of subconscious perception.[43] He conducted experiments at the Pathological Institute of the New York State Hospitals during the late 19th century, in which he showed participants cards containing a single printed letter or number at a distance where they could see only a blurred outline of either. To his surprise, when asked to guess whether the blur was a letter or number, most identified it correctly and, in some cases, could even tell him precisely what it was. Since they could not distinguish what they were seeing, Sidis concluded he must have somehow 'primed' them to respond as they did.

Since Sidis' pioneering studies, numerous researchers have confirmed his findings, emphasising the little-known powers of priming to influence people without them ever realising it. Among them was the French psychologist Alfred Binet, creator of the first IQ test. He claimed stage illusionists took advantage of their audience's natural tendency to conserve mental effort, a trait they were usually unaware of.[44]

This principle is particularly evident in priming, where magicians subtly make one choice appear easier or more appealing than others, thus influencing the audience's decisions while maintaining the illusion of free choice. It relates to a foundational idea in magic and psychology about how people make choices and process information.

There are three main aspects to this. The first is what Binet described as mental 'laziness'. The fact that people tend to follow the path of least resistance to reduce the cognitive effort involved to a minimum, even when making important decisions. In magic and marketing, manipulating this tendency might involve setting up a situation where one choice seems more natural or straightforward than others and subtly guiding the spectator towards that choice.

The second is stereotypical behaviour: people often behave in predictable ways when faced with specific situations. These common behavioural patterns are used to influence choices or actions. Finally, there are common responses, which refers to the fact that many people will give the same or similar answers when asked certain questions or presented with specific scenarios, enabling marketers to predict choices and magicians to 'read' minds. For example, when thinking of a playing card, there is a high probability that people will choose the ace of spades.[45] This is likely an example of priming since the ace of spades is usually the card on the front of the box and the first out of the pack, so it is the card people see most often.

In 2012 researcher Jay Olson, together with psychologists Alym Amlani and Ronald Rensink, confirmed this long-held belief after conducting a rigorous scientific investigation into card selection.[46] The ace of spaces emerged as overwhelmingly the most selected card, followed by the other aces. Yet despite magicians having long considered the queen of hearts to hold special status as a frequently chosen card, the research painted a more nuanced picture. The queen of hearts did show slightly higher selection rates compared to other queens, but this appeared to be driven primarily by a general preference for the hearts suit rather than anything unique about the queen itself.

Magicians, mentalists and marketers employ various methods to help plant things in spectators' or consumers' minds to create seemingly miraculous effects. Here's an example of a simple technique that gives the illusion of mind reading.

The magician asks the participant to think of something anyone could draw in less than five seconds that everyone would instantly recognise. For example, a house, the sun, a flower or a bird, though he adds: 'Please don't use the examples I've just mentioned, as that would be too obvious!' These examples, however, are given to subtly create an outdoor scene in the participant's mind, with a tree and maybe a car being the most obviously missing elements. The magician then asks the participant to think of two more drawings before finally requesting they choose the tallest object among their three choices. Due to the initial priming, the participant is likely to have thought of a tree as one of their choices, and it's expected to be the tallest touchable object. This enables the magician to apparently 'read their mind' by 'divining' that the participant is thinking of a tree.[47]

This force proves effective because it subtly guides the participant's thoughts without their awareness. The paradoxical instruction not to use the examples mentioned by the magician makes those concepts (and related ones) more prominent in the participant's mind. Additionally, giving the spectator a limited amount of time to think about what object to draw and make their drawing further limits their choice, increasing the likelihood that a tree will be chosen.

In their 2020 study, Alice Pailhès and Gustav Kuhn investigated a magic technique called the 'mental priming force', for which British illusionist Derren Brown became famous.[48]

The technique uses subtle verbal and nonverbal cues to influence a spectator's thoughts. The performer first asks participants to picture a playing card in their mind's eye, suggesting they 'make the colour bright and vivid' (subtly priming for a red card). Next, while asking participants to imagine the card, the magician forms the thumbs and fingers of both hands in the shape of a diamond. Then, as participants are asked to imagine 'little numbers low down in the corner of the card and in the top', the performer quickly traces small threes with their index finger in the air. With practice this can be performed very swiftly, typically lasting around 15 seconds, and, if successful, should prime the observer to think of the three of diamonds.

In a study by psychologists Alice Pailhès and Gustav Kuhn it was found that, when used by a skilled magician, this 'force' led to 17.8 per cent of participants choosing the three of diamonds, a significantly higher proportion than would be expected based on chance alone, and higher than in a control condition without priming.[49]

Moreover, 38.9 per cent chose a three of any suit, while 33 per cent chose a diamond of any number. When directly asked about it, all participants felt their choice to have been entirely free and remained unaware of the priming influence.

While these findings suggest that priming can work in this specific context, we should be cautious about generalising these results. First, this is a highly controlled situation – a magic performance where the audience expects some form of manipulation. This is quite different from everyday decision-making or consumer choices. Second, while the effect is statistically significant, it's still relatively small – 82 per cent of participants did not choose the target card.[50]

Although magicians are interested in using a verbal or visual prime to control what spectators see, the method's unreliability gives them pause for thought. Typically, they will have only one shot at getting it right. Due to the sheer numbers, there is far less pressure on commercial or political players. Rather than manipulating just *one* person in a specific direction, such as picking a particular card, they need to influence the most people possible to buy a certain product or vote for a particular party. For example, if used on a million people, even a 25 per cent success rate means the 'force' will still have succeeded in controlling the 'free choice' of 250,000 people.[51]

The Powers of Celebrity Priming

Advertising's diversity and widespread use make it an excellent vehicle for studying priming. In many advertisements, celebrities endorse a product or brand, seeking to persuade consumers by presenting themselves as experts.

Chanel, for example, chose actor Timothée Chalamet to be the face of its men's fragrance, Bleu, to align with the evolving concept of masculinity in the beauty industry. Being a highly influential figure with a massive following on social media, Kim Kardashian's partnership with clothing retailer SKIMS significantly contributed to the company's early success and popularity.

George Clooney's endorsement of Nespresso not only gave the coffee brand a luxury, upmarket feel, but sales increased by a third in one year alone.

Which is, of course, the reason why companies are prepared to pay huge sums for endorsements by A-list celebrities and social media influencers, whose presence is especially effective among 18- to 34-year-olds, almost a third of whom say such endorsements impact their purchases.[52]

A study by professor of business Anita Elberse and capital analyst Jeroen Verleun reported companies could expect to see an average 4 per cent increase in sales and a 0.25 per cent rise in their stock price when using a celebrity to endorse their product. For large companies, what may seem like small numbers can amount to millions, if not billions, in sales.[53]

As with the products and brands they endorse, celebrities must be seen as aspirational, likeable and trustworthy. Should they fall from grace, both the companies involved and their stars can lose millions. Former American road racing champion Lance Armstrong reportedly lost $75 million in sponsorship deals after being found guilty of doping. After being shamed by the media, world-famous athletes Tiger Woods, Michael Phelps and Michael Vick cost the brands with which they were associated, and themselves, hundreds of thousands of dollars. Next to trustworthiness, their expertise, whether genuine or claimed, is a significant component of a persuader's priming credibility.[54] The widely held belief that 'experts are usually correct' has been shown to operate as a peripheral priming cue, creating a blind spot for the truth.

'Expert celebrity-product pairings can be very successful,' says former director general of the Institute of Practitioners in Advertising Hamish Pringle. 'For example, celebrity chef Jamie Oliver's appearance in a UK food retailer advertisement was estimated to have resulted in about £313 million [US$400 million] in incremental profit over five years.[55]

A brain imaging study conducted by Vasily Klucharev and his colleagues reported that links to experts favourably increased consumer attitudes towards the items they were promoting by 12 per cent, and the probability of the consumers remembering those items by 10 per cent.[56]

Because priming occurs below conscious awareness, we often cannot appreciate how much we have been blinded to other options. In a study analogous to that conducted by Rosellina Ferraro with the bottled water brand Dasani (see Chapter 9), marketing specialists Jonah Berger and Grainne Fitzsimons primed consumers in several ways.

In one experiment, they used an approach similar to that of Esther Papies and Petra Hamstra in their dieting prime, as described in Chapter 9, to nudge university students to eat more fruits and vegetables by attaching a slogan to canteen trays. In another, they investigated how images of dogs would perceptually – or conceptually – 'prime' consumers to memories of a particular product. In this instance, trainers.

As Berger and Fitzsimons point out, 'Research has shown that when asked for the first word that comes to mind when they hear "dog", 75% of people responded with the word "cat". Consequently, we suggest that when exposed to dog images, the "cat" category will become active, and members of that category (e.g., lions, pumas) will also become more accessible.'[57]

Because Puma trainers are strongly associated, both directly (a picture of a cat adorns their logo) and indirectly (the brand name is a member of the cat category), the researchers believed the Puma brand would be more accessible in memory following dog priming.[58] This is what they found. The more often participants were exposed to dog images, the more favourably they rated Puma trainers. Similar effects were not found on evaluations of products less related to the cue (i.e. trainers from other brands).

Our cognitive blind spots, then, can be exploited; and from the subtle art of distraction to the potent influence of misinformation and priming, magicians have long understood and manipulated the quirks of human perception and cognition. And the techniques of misdirection, false evidence and psychological forcing that magicians employ find their echoes in the sophisticated strategies used by marketers, politicians and influencers in our modern world.

This exploration has revealed the double-edged nature of our cognitive processes. The same mental shortcuts that allow us to navigate a complex world efficiently can also leave us vulnerable to manipulation when skilfully exploited. However, understanding these mechanisms is not cause for despair but rather for empowerment. By recognising how our minds can be tricked, we take the first crucial step towards protecting ourselves from unwanted manipulation.

This knowledge equips us with a more discerning eye, allowing us to question our perceptions and decisions more critically.

The final chapter will build on this understanding to explore practical strategies for banishing our blind spots.

CHAPTER TWELVE
Banishing Your Blind Spots

Indian Railways had a problem. Over the past ten years, more than 72,000 people had been killed or seriously injured after misjudging the speed of fast-approaching trains.[1] With this in mind, Indian Railways made several significant trackside changes designed to make the speed of oncoming trains easier to estimate.[2] One of the first things they did was paint sleepers bright yellow, in groups of four or five, in places where frequent trespassing occurred. 'This helps trespassers judge the speed of an oncoming train by the rate at which the yellow sleepers disappear below it,' explains Rajendra Kumar Verma, the deputy general manager of Indian Railways in Mumbai.[3]

Highlighting the sleepers in this way helped trespassers judge the speed of an approaching train more accurately by increasing saliency, which as we saw in Chapter 2 plays a vital role in attracting attention and banishing blind spots. Gruesome posters depicting people being crushed to death by trains were also used to dramatically illustrate the dangers to life of crossing the tracks. 'By placing them at the exact locations where accidents were at their highest, they were able to target the audience at the most vulnerable decision-making point,' says Verma. 'Hence, they acted more carefully.' The measures cut fatalities at one station from 40 in 2009 to two in 2013 and are estimated to have reduced overall deaths and injuries by up to 75 per cent.[4]

Christopher Chabris and Daniel Simons have said that 'trying to eliminate inattentional blindness would be like asking people to try to fly by flapping their arms really rapidly'[5] – and this is true up to a point. But as the Indian Railways example shows, in many contexts blind spots are by no means inevitable. Practical steps can be taken to avoid or prevent them.

To banish blind spots from our lives and ensure we always see what matters most, we first need to identify their likely causes. Blind spots can arise in two ways: *endogenously*, as a result of the way our visual system functions; and *exogenously*, due to external factors. These, in turn, can occur unintentionally due to poor design or intentionally by those trying to influence us, such as businesses trying to reduce their costs and increase profit margins.

While many blind spots occur naturally due to our cognitive limitations, others are intentionally created to manipulate our perceptions and decision-making. This is where the concept of 'illusioneering' comes into play. Just as Indian Railways used visual cues to overcome natural blind spots, some entities use sophisticated techniques to create artificial ones.

The Hidden Cost of Illusioneering

Illusioneering is the deliberate and systematic exploitation of cognitive biases and perceptual limitations to create or amplify psychological blind spots, typically for commercial, political or strategic gain. It involves carefully designing information environments, user experiences or communication strategies that exploit known cognitive shortcuts and decision-making heuristics.

For example, the digital architecture of online platforms can be engineered to shape user behaviour, with the ultimate aim of influencing how people think, feel and act while ensuring consumers remain unaware of these manipulations.

Have you ever noticed, for example, that on opening a social media app you often find the feed's bottom post is only partially visible, an unspoken instruction from the software designers and engineers who built the platform to 'Scroll down!'

'This', explains Michael Bossetta, 'will expose you to the advertisements, which helps pay their wages while subtly shaping user engagement.'[6]

YouTube's recommendation system, at the heart of the platform's design, for example, exerts a tremendous influence over the viewing and consumption habits of its 2.49 billion monthly active users, and more than 80 million paid subscribers. By prioritising engagement, it has created an environment criticised for leading users towards more extreme content.[7]

As explained in Chapter 11, those we termed 'Fun Junkies' have become so reliant on platform algorithms for news that they disengage from news-seeking altogether.

Illusioneering introduces new areas of cognitive blindness or magnifies existing ones by exploiting innate mental shortcuts. These include a tendency to focus on information that supports already-held beliefs, expectations or hypotheses (confirmation bias); overreliance on an initial piece of information to make subsequent judgements (anchoring bias); and loss aversion, in which consumers choose the option they regard as safer when offered the choice between a high-payoff but riskier option and a safer, low-payoff option. While often associated with marketing and with digital platforms, illusioneering can be employed in various fields, including politics, education and social engineering.

Illusioneering helps ensure consumers are continually paying top dollar for services that are gradually deteriorating in quality, often without realising they are paying more while getting less. 'In today's world, brands and labels hold immense power', points out Priyanshu Sharma, a sales and marketing specialist at Tejraj Group, a construction and real estate company 'weaving a narrative of perceived value that often overshadows the true essence of products. This begs the question: Are we, as consumers, chasing shadows instead of substance?'[8]

Here are nine examples of how both online and bricks-and-mortar companies create an illusion of value while gradually eroding their offerings' actual quality or quantity might work in practice. As we explore them, consider how you might have encountered similar tactics in your experiences with products, services or information platforms.

1. *Confirmation Bias* Tailoring news feeds or search results to show information that aligns with a user's existing beliefs. Example: A social media platform algorithm that predominantly shows political content matching a user's ideology, reinforcing their views and creating an echo chamber.

2. *Anchoring Effect* Presenting an initial piece of information to influence subsequent judgements. Example: An online retailer showing a high 'original' price crossed out next to a 'sale' price, creating an anchoring effect and leading customers to perceive a better deal than may exist.

3. *Availability Heuristic* Making certain information more readily available to influence the perception of frequency or importance. Example: A news website prominently features stories about rare but sensational crimes, leading readers to overestimate the prevalence of such events.

4. *Framing Effect* Presenting the same information differently to elicit different responses. Example: A political campaign describing a policy as '95 per cent effective' rather than 'fails 5 per cent of the time' to garner more support.

5. *Sunk Cost Fallacy* Encouraging continued investment in a course of action due to past investment. Example: A mobile game design that makes players feel they've invested too much time to quit, even if they're no longer enjoying it.

6. *Bandwagon Effect* Highlighting the popularity of a product or idea to encourage others to adopt it. Example: A restaurant app showing 'most popular' dishes, influencing diners' choices based on others' preferences.

7. *Loss Aversion* Framing choices in terms of potential losses rather than gains. Example: An insurance company advertisement emphasises the risks and potential losses of being uninsured rather than the benefits of insurance.

8. *Choice Overload* Presenting an overwhelming number of options to paradoxically reduce thoughtful decision-making. Example: A streaming service offering thousands of titles, making it difficult for users to choose and often defaulting to suggested content.

Social Media Scams

Social media platforms, for some of the reasons described earlier in this chapter, have become breeding grounds for sophisticated illusioneering scams that exploit our trust in familiar interfaces, our desire for recognition or financial gain, and our tendency to overlook subtle anomalies when we're engaged in what seems to be a routine interaction.

One particularly insidious form of this exploitation targets content creators – individuals who, despite being tech-savvy, fall victim to carefully crafted deceptions. The scam has gained traction since 2023 and preys on content creators by falsely presenting themselves as a legitimate business opportunity. It demonstrates how our cognitive biases can be weaponised against us, even when we believe we're operating in a familiar, safe environment.

Masquerading as podcast managers or representatives of reputable brands, the scammers carefully select their targets – content creators with substantial followings. The initial contact appears legitimate, often featuring a generous payment offer for appearing on the podcast.

For many content creators, podcast invitations are a regular occurrence, making the approach credible. The scammers exploit this familiarity by arranging a briefing call to discuss the podcast details and payment methods. During the call, they claim that the podcast will be recorded via a Facebook Live event on the creator's public page, citing a sponsorship from a well-known brand as the reason behind this unconventional setup.

At this point, the call takes a treacherous turn. The scammer requests that the creator share their screen, guiding them through a series of settings under the guise of setting up the joint Facebook Live event. In reality, they are attempting to gain control of the creator's page through the back end of Meta. If successful, the scammer seizes control, removing original content and repurposing it, such as advertising fraudulent products.

They leverage the credibility of the hacked account, reaching out to other content creators via direct messages and perpetuating the cycle of deceit. This ripple effect makes the scam particularly dangerous as it quickly expands its reach, ensnaring more victims. Since 2023, the fraud has left a trail of hijacked accounts and bewildered victims.[9]

Professional athlete and fitness coach Gabbi Tuft describes how, after her appearance on a podcast had been scheduled, the scammer requested a pre-interview meeting to ensure everything ran smoothly. During this, she was persuaded to change her Facebook settings. Within hours, she found herself locked out by the scammers, who were now in complete control of her account. 'I am deeply disappointed and disturbed by this deceitful act whereby individuals are capitalising off the name recognition of one of the most high-profile celebrities and podcasts right now,' said an aggrieved Tuft. 'It's unfortunate that individuals would exploit the trust and goodwill of others for personal gain'.[10]

Scammers have targeted fans to hijack the accounts of such celebrities as the American singers and musicians Dolly Parton, Bret Michaels and Blake Shelton and then make posts as if they were the holders themselves.

To protect ourselves from illusioneering, we need to cultivate awareness and critical thinking. Some useful techniques include questioning gradual changes to products and services we use regularly; comparing value rather than just price; conducting thorough research before upgrading to a premium product or service to assess whether the added features justify the cost, taking our time before reaching a decision and making sure to avoid doing so when tired, angry or under stress. This trinity of physical impairments significantly increase our vulnerability to illusioneering scams.

The Three 'Fs' of Blind Spots Creation

When fatigued, frightened or furious, we become far more vulnerable to failing to see everything we need. Sometimes, we intentionally inflict these situations on ourselves by burning the midnight oil, being frightened for fun, or allowing ourselves to lose our temper. Others can also intentionally inflict them on us to encourage our blind spots.

Creating anger and fear, on occasions intense hatred and terror, has never been easier thanks to social media, which enables negative messages to spread faster and further than at any time in human history[11] and enable anger to spread contagiously.[12]

Over the last few years, social media platform designers have admitted that their systems are addictive and exploit negative 'triggers'[13] and rip apart the social fabric by hijacking users' minds.[14] As a recent U.N. report pointed out: 'Online hate, with the speed and reach of its dissemination, can incite grave offline harm and nearly always aims to silence others'.[15]

The hatred-based violence of recent years, together with the fear such actions generate, is neither random nor anomalous but the inevitable consequences of inhabiting spaces where racist, sexist, anti-Black, anti-Muslim and anti-Semitic views are normalised. Spaces that have provided a means by which emotions and actions may be manipulated for commercial and political purposes.

For an example of the impact fatigue can have, consider the experience of one of the authors (David), who some years ago spent several weeks at sea studying the working lives of Britain's trawlermen, during which time he encountered his first, and almost his last, life-threatening blind spot. For days, the ocean had been what sailors term 'lumpy', with wind force measuring between 3 and 4. Uncomfortable, but hardly alarming. Fifteen days into the trip, a weather forecast warned that a Force 9 gale (in the US, a 'Whole Gale') was imminent.

The wind grew ever stronger over the next few hours, and the seas rougher. Great patches of foam, driven across the heaving waters by 90km/h (55mph) winds, turned the thundering waters a ghostly white and as rugged as snow-covered mountains. Just after midnight, reluctant to remain below deck, David offered to take a mug of strong, black coffee to the helmsman, Jack. Alone in the pitching wheelhouse, Jack had been battling for hours to keep the trawler bow-on to the mountainous waves.

Jack was still on the bridge a couple of hours later when he noticed the lights of another vessel heading directly towards them. Seemingly unconcerned by an impending collision, he made no move to alter course. Only at the last moment, when the other trawler was almost on top of their vessel, did he swing the helm hard to starboard. There was the agonising grinding of metal as two boats scraped past one another. Then, as suddenly as it had appeared, the other ship had vanished in the blackness. All aboard attributed their brush with death to fatigue. In this, they were almost certainly correct.

Research has shown that the brain regions controlling alertness produce increased sleepiness and diminished mental capacity between 2 a.m. and 7 a.m.[16] The risks of mistakes due to blind spots become far greater during this period, with decisions and solutions often riddled with errors, as the following tragic examples demonstrate.

In January 1986, the seven-person crew aboard the Challenger Space Shuttle perished when the rocket exploded after liquid oxygen was mistakenly drained from the shuttle's external tank shortly before launch. Fatigue was reported to be 'one of the major factors contributing to this incident'. The operators, who had been on duty for 11 hours, were on the third day of working 12-hour shifts.[17]

The fatigue-induced blind spots leading to the Challenger disaster are by no means rare. The worst accident in US commercial nuclear power plant history occurred at 4 a.m. on 28 March 1979, after workers failed to spot the loss of core coolant water from a stuck valve. As a result, Reactor 2 at the Three Mile Island plant in Pennsylvania spewed radioactive gases and iodine into the surrounding countryside.[18]

Medical errors also rise significantly during the early hours, when staff are often overly tired.[19] 'The increase in attentional lapses and decreased subjective alertness that night-shift workers experience, especially on the first day of the night shift, lead to an impaired ability to maintain focus,' reports MD Ann Rogers. 'A study of hospital staff nurses showed that 47 per cent of episodes of drowsiness and 29 per cent of actual sleep episodes occurred between 00:00 and 06:00. Drowsiness and falling asleep on duty are associated with significantly more risk of making an error.'[20]

On the roads, while up to 90 per cent of vehicles are parked up overnight, 29 per cent of accidents occurred during darkness, with almost half of them involving a fatality.[21]

Rarely recognised or appreciated is the blind-spot-generating nature of present-day society, with its artificial light, caffeine use and late-night screen time. 'Insufficient sleep (short sleep duration and poor sleep quality) has become pervasive in modern societies with 24/7 availability of commodities,' says sleep expert Jean-Phillipe Chaput. 'Factors responsible for this secular decline in sleep duration are numerous and generally ascribed to the modern way of living'.[22]

In one study, the authors invited young men and women to participate in what they believed to be team-building exercises.[23] Because an early start was needed, they were checked into a hotel the night before and divided into Blue and Yellow teams.

At one in the morning, a researcher knocked on the Yellow Team's doors and told the bleary-eyed occupants to tackle a tough IQ test. This took about 30 minutes, after which they resumed their disturbed sleep. Two hours later, we again roused them from their slumbers to complete a second IQ test. Once finished, they went back to bed and slept until 7.30, when both groups had to get up and come down to breakfast.

Considering the time they had spent struggling with the IQ tests, we calculated that, on average, the eight hours of sleep enjoyed by the Blue Team had been slashed to five for members of the Yellow Team. During the morning, the two teams competed against one another, playing games that they believed were the purpose of the study.

At lunchtime, team members were given money to buy food at a nearby supermarket. While Blue Team members considered the options more carefully and generally sought a balanced diet, healthy eating options went unseen by the sleep-deprived Yellow Team, all of whom chose foods higher in sugar and fat. By depleting self-control, sleep deprivation significantly influences what is seen or unseen over a wide range of purchases.

In another study, conducted at Brands Hatch racetrack, we kept a group of motorists, including champion rally driver Jason Plato, awake for 24 hours, testing their mental and physical abilities every couple of hours, using a car simulator to motor along busy city streets and down narrow country lanes. By comparing their performance across time, the consequences of sleep deprivation on their coordination and decision-making became all too apparent. Even professional rally driver Jason found himself driving on to curbs, narrowly avoiding collisions and almost hitting pedestrians.[24]

When sleep is being determined by external forces, such as an overly demanding boss who expects employees to be available at all hours of the day and night, do your utmost to ensure you enjoy an average of 50 hours' uninterrupted sleep per week. While some claim to 'get by' perfectly well with as little as three or four hours of sleep per night, research suggests the only way to eliminate fatigue-induced blind spots is to ensure you have between seven and eight hours' uninterrupted sleep each night.

An intriguing experiment demonstrated the importance of a good night's sleep before making an important decision. Psychologist Alyssa Sinclair and her colleagues[25] recruited volunteers to participate in an imaginary online garage sale. They searched through virtual boxes of unwanted goods for items to include in the sale. While most of the items they found, an old alarm clock or a potted plant, could be described as junk, having little value, a few, like a nice lamp or a teddy bear, were 'gems' and worth far more.

Participants were motivated to determine which boxes were most valuable since they earned real cash based on which boxes they chose. Unbeknown to them, the total value of items was the same in each box; only the ratio of 'junk' versus 'gems' varied.

In some, all the valuable items were on top, and participants spotted them immediately. In others, gems and junk were intermixed, or items were clustered in the middle or bottom. After opening the boxes, participants were asked to estimate the value of them and choose their favourite items. Some participants judged the boxes immediately, but others 'slept on it', making up their minds the following day.

When they had to make an immediate decision, participants tended to remember and judge boxes by the first few items they encountered, rather than all the contents. When gems were seen before low-priced items, participants were likelier to pick that box than if they had seen the cheap stuff first. In such cases they also tended to estimate the box's value at about 10 per cent more than it was worth.

When forming an opinion, participants who made up their minds immediately were strongly influenced by first impressions, a psychological phenomenon known as the 'primacy effect'. Those who enjoyed a night's sleep made more rational choices, equally favouring boxes with valuable items on the top, in the middle, or at the bottom.

'This is an exciting first look at how our brains summarise a rewarding experience,' says Alison Adcock, in whose laboratory the 'garage sale' study was conducted. 'When it's over, our brain knits it all together in memory to help us make better choices – and that neat trick happens overnight.'[26]

Here are six practical strategies for improving your night's sleep.

1. Exercise with care. Studies show adults who exercise for at least 30 minutes a day sleep an average of 15 minutes longer than those who do not exercise. Physical activity can also help to reduce sleep disorders, such as insomnia, daytime sleepiness, and sleep apnoea. This does not necessarily mean putting on your jogging shorts and going for a long run. If this is what you enjoy, then, of course, do so, but even a stroll around your neighbourhood is better than nothing. That said, you should avoid high-intensity exercise such as interval training in the two hours before you go to bed.[27]

2. Wind down gradually by establishing a pre-bedtime routine you consistently follow.

3. Do not drink coffee, tea or caffeinated cola drinks in the six hours before going to bed. Especially avoid high-caffeine energy drinks and premium coffee, whose consumption has significantly increased in younger age groups and is associated with chronic sleep restrictions.[28] Stop drinking alcohol four hours before bed since it can disrupt sleep patterns and make it harder to fall into a deep sleep. Try one or more of the following instead.

- Warm milk: contains tryptophan, a precursor to melatonin, the hormone that helps you feel sleepy.
- Chamomile and valerian tea: known for their calming properties.

- Tart cherry juice (also known as sour cherry): contains melatonin and may have an antioxidant effect that helps with sleep.

4. When in bed, avoid using a computer, tablet, or mobile phone immediately before turning off the light. Many such devices emit a lot of blue light, encouraging wakefulness and increasing the risk of longer-term damage to one's health. This light affects your body's capacity for creating melatonin while interrupting the circadian rhythm; such interruptions play a role in the development of heart disease, cancer, Type 2 diabetes and cognitive dysfunctions.[29] Switch digital products off at least two hours before you want to settle down and sleep for the night.

5. Turn off or cover any device, such as a clock, that emits even the faintest illumination. Ensure the curtains are thick and light-proof, especially in a town or city where streetlights may shine through your bedroom windows at night. The darker your bedroom, the better you will sleep.

6. Ensure that while your bed is comfortably warm, the room never becomes overheated and stuffy. A slightly humid room is also better than an overly dry one since it prevents your sinuses from drying out at night, which can restrict your breathing and make it harder to enjoy restful sleep. If noise is not a problem, open a window to ensure fresh air flow.

You can find a video about enjoying a good night's sleep by going to lewisandleyser.com/sleep or scanning the QR code.

When we're tired, our cognitive resources are depleted and we have a diminished ability to maintain attention, making us more susceptible to illusioneering tactics. Tired individuals are more likely to rely on mental shortcuts that others can exploit to guide decisions in their favour. Fatigue can erode self-control, making us more vulnerable to impulse purchases or hasty decisions. A fatigued consumer browsing an e-commerce site late at night might, for example, be less likely to notice that a 'limited-time offer' has been running for weeks or that a product's size has been subtly reduced while the price remains the same.[30]

Alongside fatigue, another of the 'f's that can have a serious impact on our susceptibility to blind spots is fear. Setting up bogeymen that arouse fear has long been the surest and fastest way of distracting people from noticing situations or actions those in power would prefer remain unseen.[31]

As political theorist Corey Robin points out, far from being lone individuals' work, many fear-inducing political stories and conspiracy theories are the products of highly organised and well-funded organisations. 'These activities,' says Robin, 'are neither spontaneous nor episodic: they require the ongoing labour of elites and collaborators ... to be sustained over time, men and women must be hired, paid, supervised, and promoted.'[32]

Because it would be unethical to create situations in which naïve subjects became genuinely fearful, our research has focused mainly, but not exclusively, on 'fun fear'. That is the self-imposed terror people experience when screaming their heads off on a white-knuckle theme park ride. The only difference is the mental label people attach to it. Rather than telling themselves, 'I'm terrified', participants view the sensations as indicating excitement and enjoyment. Sensors measuring heart rate and skin conductance have shown that many responses present when a person is genuinely afraid, such as increased heart rate, dry mouth, sweating and mental confusion, are also found in 'fun fear'.

Participants wore eye-tracking glasses to record their gaze direction and what they saw while riding a roller-coaster. The glasses showed that their attention was restricted as fun fear mounted, leaving them incapable of seeing anything outside an increasingly narrow field of vision. You can watch one of our white-knuckle ride studies by going to lewisandleyser.com/roller-coaster or scanning the QR code.

Dr Daniel Parisi confirmed our findings in an experiment conducted during the famous 'Running of the Bulls' in Pamplona, north-eastern Spain. Since the festival started in 1910, 16 runners have been gored to death and hundreds injured, struck by the bulls or trampled by the panic-stricken crowd. 'Although it's a voluntary and somewhat contrived danger,' Parisi explains, 'the people's flight is real enough to offer a rare insight into how humans behave as they try to escape danger.'[33]

Parisi and his colleagues set up cameras above the Pamplona streets to record a bird's-eye view of the event. Fear spread like wildfire through the densely packed runners as the bulls approached. With all their attention on the charging beasts, anything else, including other runners, remained unseen. People began colliding with one another and falling heavily on to the cobblestoned streets.

'In a laboratory, this kind of experiment with real danger is unethical. But this festival is almost like laboratory conditions,' explains Dr Parisi. 'The street is easy to monitor, the runs happen several times and in the same place, and the stimulus

– the sudden appearance of a charging bull – is a consistent motivational force.'

As psychiatrist David Hawkins comments: 'Fear is the favoured official tool for control by oppressive totalitarian agencies and insecurity is the stock-in-trade of major manipulators of the marketplace. The media and advertisers play to Fear to increase market share ... Fearful thinking can balloon into paranoia ... and because it's contagious, become a dominant social trend.'[34]

Banishing fear-induced blind spots requires people to step back from what is scary and take a more relaxed and objective view of what is happening. We should distance ourselves from sources of fear and ignore scaremongering conspiracy theories and unsubstantiated rumours.

By narrowing our focus, fear can create a sense of urgency that illusioneers exploit to push for quick decisions without thorough consideration. When afraid, we tend to overestimate risks, which can be manipulated by those looking to sell unnecessary products or services. Fear also activates the emotional centres in our brain, potentially bypassing more logical decision-making processes. For instance, an insurance company might use fear-inducing marketing that exaggerates unlikely scenarios, creating a blind spot around the probability of needing extensive coverage. This could lead consumers to purchase more insurance than necessary.

If you would like to understand the link between anxiety, fear and the way we breathe, go to lewisandleyser.com/breathing or scan the QR code.

If you would like to better understand the link between the rate at which you breathe and your level of sympathetic arousal, go to lewisandleyser.com/biofeedback or scan the following QR code.

Just as fear narrows people's focus and blinds them, perceptually and conceptually, to everything but escaping the source of their terror, so too does rage cause them only to see and think about a tiny portion of their surroundings.[35] As with genuine fear, anger is a complex emotion to study experimentally. A few years ago, a British television company invited one of us (David) to act as a consultant on a programme entitled *Red Mist*. A group of self-identified angry men, recruited from all over the UK, were invited to attend an audition in a central London studio. Each was warned arriving late would result in the audition being cancelled and, with it, any chance of their appearing on TV.

All arrived promptly and were directed to a waiting room staffed by a surly and unhelpful security guard. As the time ticked by, first towards their scheduled appointment and then past it, they became first irritated, then angry and finally – in at least one case, blind with rage.

What they failed to notice, although there were plenty of visual pointers to the truth, was they were already taking part in the programme. The 'waiting' room was equipped with studio lights and miniature, but still clearly visible, TV cameras. The 'security guard' was an actor. Both he and the room were wired for sound.

Instead of seeing the whole room with its cameras and microphones, they saw only the unfortunate object of their rage: the actor playing a truculent and deliberately time-wasting security man.

When overwhelmed by either fear or anger, an effective way of banishing emotion-induced blind spots is by withdrawing from the situation until that powerful emotion is under control. Remember that, often, anger is deliberately triggered by those who want you to dance to their tune. When you lose control, you fail to see whatever it is they would sooner you didn't.[36]

One ancient technique for managing anger is to write angry thoughts on small pieces of paper before setting them alight and letting them fall into a water-filled bowl. The value of this ancient technique was demonstrated in a study by psychologists Yuta Kanaya and Nobuyuki Kawai.[37] After insulting their subjects, they asked them to write down these provocations and how they had made them feel. They were then told to toss the paper into a waste bin or keep it on their desks. A second group was instructed to shred the document or place it in a plastic box. Kanaya and Kawai reported that while the anger of those discarding or shredding the insults subsided, participants who retained a hard copy of the insult experienced only a slight decrease in their overall anger.

Their findings may explain the catharsis offered by the annual Japanese festival of *Hakidashisara*, at which people smash small discs representing things that make them angry (*hakidashi* refers to the purging or spitting out of something, and *sara* to a dish or plate).

By understanding how our emotions are linked to blind spots, we can be more vigilant when tired, afraid or angry. Recognising these states as periods of increased vulnerability to manipulation or bad decisions can help us pause, take a step back and make more considered decisions. This awareness is crucial in protecting ourselves from manipulative practices that exploit our cognitive limitations.[38]

Blind Spots and Magical Thinking

Magical thinking is the psychological phenomenon of believing we can influence outcomes by means without a logical connection to the situation, such as influencing our luck by touching wood, avoiding walking under ladders or carrying a lucky charm. While dismissing such actions as silly superstitions may be tempting, the truth is that magical thinking is a deeply ingrained part of the human experience.

Our brains are hardwired to seek patterns and explanations, which was vital for the survival of our prehistoric ancestors. Magicians often capitalise on this pattern-recognition process when crafting their illusions. They understand that our brains instinctively filter out anything deemed irrelevant, and they use this knowledge to hack into our brains as that occurs, to create seemingly impossible feats.[39]

Awareness of magical thinking teaches us the importance of maintaining critical thought throughout our lives. As sharp and nuanced thinkers, we must stay vigilant about the factors that make us vulnerable to magical thinking. We should continually question the origins of our beliefs and acknowledge that our rational minds are not the sole influence on our thoughts. Our convictions can be moulded by the lessons our parents instilled in us from childhood, the things we wish were true and the conclusions our experiences lead us to draw.[40]

Magical thinking can also create obstacles to our putting in the necessary work to reach our objectives. Many of us believe in the power of visualisation as a helpful technique. However, we must be careful not to fall into the illusion that simply conjuring a vivid mental image of our desired outcome is enough to bring it to fruition and not do the work involved.

Banishing Blind Spots in Everyday Life

We each inhabit our own unique *Umwelt*. This applies not only to individuals but also to corporations, political parties and social or religious organisations, where the prevailing *Umwelt* is usually that of the leader, whether company chairman, president or influencer.

The fact others see what is unseen by us does not mean we are right and they are wrong – nor, of course, that they are right and we are wrong. In a social media- and AI-dominated world, opportunities for manipulating blind spots have never been easier or more extensively employed. Awareness of conceptual blind spots and biases in everyday life requires self-reflection and an openness to challenging one's assumptions. The following seven strategies can help do just that.

1. Broaden your worldview by exposing yourself to a wide range of viewpoints, opinions and experiences.

2. Regularly question your beliefs and assumptions. Ask why particular views are held and whether they are based on evidence or preconceived notions. Establishing a habit of self-enquiry can reveal your unconscious biases.

3. Whenever possible, allow yourself time to review every important choice or decision you make to ensure alternatives have been carefully considered.

4. Challenge stereotypes as a proactive step to understanding cognitive biases and common conceptual blind spots. Reflect on your emotional responses to various situations that often indicate underlying biases.

5. Make a determined effort to understand and relate to others' experiences and emotions; this can help bridge gaps in understanding and reveal previously unnoticed perspectives.

Empathy can be developed through active listening, perspective-taking exercises and immersing oneself in diverse social situations.

6. Gain new insights by stepping outside familiar thought patterns and embracing discomfort. Intentionally exposing yourself to situations, people and ideas outside your comfort zone can help identify blind spots.

7. Avoid perceiving situations as good or bad, right or wrong, honest or deceptive, white or black, 'yes or no', or 'either/or'. In an increasingly complex and interconnected world, binary options are rare. We can only view a situation objectively from a different perspective.

In our journey through the landscape of cognitive blind spots, we've explored how our minds naturally filter and interpret the world around us, often leaving crucial information in the shadows.

At an average reading distance, Figure 83 is a picture of Albert Einstein. Viewing the same image from a distance transforms it into Marilyn Monroe.

Figure 83

If you have difficulty making the switch, you can watch a video of the transformation by going to lewisandleyser.com/einstein-marilyn or scanning the QR code.

There is no right or wrong answer as to what the picture shows. It's simply how you choose to see it.

By understanding the visual processes explored in this book, we provide ourselves with the tools to see beyond our initial perceptions, question our assumptions and seek perspectives that challenge our preconceptions. By doing so, we expand our awareness, make more informed decisions, and navigate our complex world more clearly.

In life, as in magic, there is always more than meets the eye – and true wisdom lies in constantly striving to unveil the previously *unseen*.

Acknowledgements

Thanks to Sarah Lambert, our brilliant and insightful Bloomsbury editor, without whom this book would never have been written. We also thank Dr David Groom, Peter Hughman, Steven Matthews, Roger Bromiley, Angela Carriero, Gina and Thomas Beresford, Pui Hang Yip, Dr Mark Radon FRCR, consultant neuroradiologist; Jonathan Hyder, BDS, LDSRCS (Eng); Duncan Smith, Mindlab's MD; Joe Hilling, Mindlab's director of innovation; and Juliane Beard, Mindlab's director of research.

Image credits: The Bonnie Prince Charlie illusion (Figure 29) is used with permission from Vanessa Martin, curator at the West Highland Museum. The 'butterfly' brain scan (Figure 43) appears courtesy of Frank Gaillard, founder of Radiopaedia. org. rID: 2589.

Notes

Introduction

1. Melton, H. K. & Wallace, R. (2010). *The Official C.I.A. Manual of Trickery and Deception*. Hardie Grant; Robinson, B. (2008). *The MagiCIAn*. Lybrary.com.

Chapter 1: Looking Without Seeing

1. Fiandaca, C. (2022) New Boston Police Commissioner Michael Cox was beaten by fellow officers in 1995. *WBZ News*.
2. Conley's sentence was overturned in 2005. Not because anyone now believed he was telling the truth but because the prosecution had failed to reveal an FBI memo casting doubt on the credibility of one of the government's witnesses. On 19 May 2006, more than 11 years after the original incident, he was reinstated as a Boston police officer, but only after being obliged to repeat, at age 37, the same training he had undertaken as a recruit. He was granted $647,000 in back pay for the eight years he was out of the force, and in 2007 he was promoted to detective sergeant. He died after a lengthy illness in 2024. After a prolonged recovery, Michael Cox returned to the force for the next thirty years before finally being promoted to commissioner.
3. Simons, D.J. & Chabris, C.F. (1999). Gorillas in Our Midst: Sustained Inattentional Blindness for Dynamic Events. *Perception* 28 (9): pp. 1059–1074.
4. Neisser, U. (1976). *Cognition and Reality: Principals and Implications of Cognitive Psychology*. San Francisco: W. H. Freeman.
5. Ho, V. (2013). Absorbed device users oblivious to danger. *The San Francisco Chronicle*, 7 October. Retrieved from www.sfgate.com
6. Hyman. I. E. (2016). Unaware Observers: The Impact of Inattentional Blindness on Walkers, Drivers, and Eyewitnesses. *Journal of Applied Research in Memory and Cognition*: pp. 1–20.

7   Edkins, G. (2016). Human Factors, Human Error, and the role of bad luck in Incident Investigations, Safety Wise; Wilson, L. (2023). Human error is not a cause it's a consequence. *Industrial Safety Review.*

8   Masunaga, S. (2024). Boeing 737 Max 9 plane that lost door plug was missing bolts, NTSB says. *Los Angeles Times*, 6 Feb. The Max 9 landed safely, with the only 'casualty' being a teenage passenger seated close to the gaping hole, whose shirt was ripped from his back by the explosive expulsion of cabin air.

9   Binet, A. (1894). Psychology of prestidigitation. *Annual Report of the Board of Regents of the Smithsonian Institution.* Washington, DC: Government Printing Office: pp. 555–571. One of the conjurors present, Georges Méliès, used the same approach when, not long after, he invented the early cinema's first special effects.

10 Those taking his tests included 14-year-old Helen Keller, who, despite being born blind and deaf, became a world-famous author, and the serial killer H. H. Holmes. Over the years, Holmes and his accomplice are believed to have murdered at least 200 people. Arrested in Boston in 1896, Holmes was tried, found guilty and hanged.

11 Jastrow, J. (1896). Psychological notes upon sleight-of-hand experts, *Science* 3: pp. 685–689.

12 Kuhn, G., Amlani, A. A. & Rensink, R. (2008). Towards a science of magic. *Trends in Cognitive Sciences*, vol. 12 no. 9: pp. 349–354.

13 Evans, E. & Craver, N. (2003). *The Secret Art of Magic.* Magic to Laugh By.

14 Despite the Cartesian implications, we have opted to use personal pronouns when discussing the brain since this is how people typically describe this and every other organ of the body, such as the heart, the leg, the kidneys, etc. It is, however, important to recognise that there is no 'creature' in our head who claims ownership of our body and brain.

15 While human brains have a significantly faster processing speed than those of most other animals, limits to their physical structure mean this cannot be significantly increased. However, they can be improved in various ways, such as through regular exercise, proper nutrition, and mental exercises.

16 Cartwright-Finch, U. (2005). Inattentional Blindness: the role of perceptual load, effects of stimulus type and position, and

development over childhood. PhD Thesis, University College London.
17 Aristotle (1955). *On dreams*. In: *Aristotle. Parva naturalia*, ed. W. D. Ross. Clarendon Press: pp. 175–223.
18 Quintilianus, Marcus Fabius (1920) [c. 95]. Institutio Oratoria, trans. H. E. Butler. Loeb Classical Library. Cambridge, MA: Harvard University Press.
19 Hankins, Thomas L. (1980). *William Rowan Hamilton*. Baltimore and London: Johns Hopkins University Press.
20 Kolers, P. A. (1969). Voluntary attention switching between foresight and hindsight. *Quarterly Progress Reports*: 92, 381-385. Research Laboratory of Electronics, MIT.
21 Neisser, U. & Becker, R. (1975). Selective looking: Attending to visually specified events. *Cognitive Psychology* 7: pp. 480–494. Their idea was based on work done during the Second World War by American psychologist James Jerome Gibson. While serving in the US Army Air Forces between 1942 and 1946, James Gibson was involved with groundbreaking research on aircraft identification. After the war, he went to Cornell University, where he developed an 'ecological approach' to studying visual perception. By contradicting 'a centuries-old model of the origins of human knowledge', Gibson's radical new theory completely changed how psychologists thought about how we make visual sense of our surroundings.
22 Becklen, R. & Cervone, D. (1983). Selective looking and the noticing of unexpected events. *Memory & Cognition* 11: pp. 601–608. Other researchers had conducted similar studies around this time, including: Becklen, R., Neisser, U. & Littman, D. The effect of event similarity on selective looking. Manuscript in preparation, 1983; Neisser, U. & Dube, E. F. Interrupting the perceptual cycle: When do we notice unexpected events? Paper presented at the meeting of the Eastern Psychological Association, Washington, DC, March 1978; Neisser, U. & Rooney, P. Noticing unexpected events in selective looking: A new criterion. Unpublished manuscript, 1982; Becklen, R. A method for determining per cent on-target time in continuous attention experiments with varying target and response probabilities. Unpublished manuscript, 1983.
23 Released on YouTube in 2010, their 'Invisible Gorilla' video clocked up 26 million views. Once one knows what to look

out for, the gorilla immediately becomes apparent, and you are left wondering how anyone could fail to see the chest-thumping beast. To watch the video go to: www.youtube.com/watch?v=vJG698U2Mvo

24 Miller, G. A. (1956). The magical number seven, plus or minus two: Some limits on our capacity for processing information. *Psychological Review*, vol. 63: pp. 81–97.
25 Sperling, G. (1960). The information available in brief visual presentations. *Psychological Monographs* 74 (whole no. 498).
26 Cowan, N., Morey, C. C. & Chen, Z. (2007). The legend of the magical number seven. In S. Della Sala (ed.), *Tall Tales About the Brain*. Oxford University Press.
27 Martens, M.H. (2000). Assessing Road Sign Perception: A Methodological Review. *Transportation Human Factors* 2 (4): pp. 347–357.
28 Read, D. (1983). Detection of Fs in a Single Statement: The Role of Phonetic Recoding. *Memory & Cognition* 11 (4): pp. 390–399. Having to count down from 100 while reading the sentence divided your attention, making it more likely you would overlook the 'f' in 'of'.
29 Ptak, R. (2011). The Frontoparietal Attention Network of the Human Brain: Action, Saliency, and a Priority Map of the Environment. *The Neuroscientist:* pp. 1–14.
30 Banks, J. (1998) *Endeavour Journal of Joseph Banks: The Australian Journey*. HarperCollins.
31 Simons, D. J. (1999). To see but not to see: Review of 'Inattentional Blindness' by A. Mack and I. Rock (1998). *Journal of Mathematical Psychology*, vol. 43: pp. 165–171.
32 Helmholtz as quoted in James, W. (1890/1950). *The Principles of Psychology*. Volume 1. New York: Dover.
33 White, A. L. & Carrasco, M. (2011). Feature-based attention involuntarily and simultaneously improves visual performance across locations. *Journal of Vision* 11 (6): 15: pp. 1–10.
34 Wolfe, J. M. (2000). The deployment of visual attention. In *RTO Meeting Proceedings 45: Search and Target Acquisition*. NATO-RTO, Utrecht, Netherlands.
35 Exner, S. (1882). Archiv für Anatomie, Physiologie und wissenschaftliche Medicin, vol. 28: p. 487.
36 Sengupta, B. & Stemmler, M. B. (2014). Power Consumption During Neuronal Computation. In *Proceedings of the IEEE*, vol. 102 (5): pp. 738–750.

37 Lavie, N. & Fockert, J. de (2006). Frontal control of attentional capture in visual search, *Visual Cognition* 14 (4/5/6/7/8): pp. 863–876.
38 Treat, J. R. (1980). A study of precrash factors involved in traffic accidents. HSRI Research.
39 Macdonald, J. S. P. & Lavie, N. (2008). Load Induced Blindness, *Journal of Experimental Psychology* 34 (5).
40 Mackinnon, I. (1994). Crash driver 'had glasses in hand': Pathologist and survivors give evidence on last moment of victims in a school minibus accident. *The Independent*
41 Finlay, M. (2023). The various factors that downed United Airlines Flight 173 on this day in 1978. *Simple Flying*.
42 National Transportation Safety Board (2005), Marine Accident Report, Washington, DC.
43 Kinsey, M.J., Gwynne, S.M., Kuligowski, E.D., et al. (2019). Cognitive biases within decision making during fire evacuations. *Fire Technology* 55 (2): pp. 465–485.

Chapter 2: Now You See It – Now You Don't

1 Simons, D. J. & Levin, D. T. (1998). Failure to detect changes to people in a real-world interaction. *Psychonomic Bulletin and Review* 5 (4): pp. 644–649.
2 The flicker paradigm involves alternating between two changing versions of a visual scene. This method has been particularly useful in studying change blindness and attentional processes. The flicker paradigm also provides insights into the interaction between short-term memory and perception, as individuals need to retain information from the initial scene to detect changes. Market researchers can use it to measure how strongly different products draw attention in a busy shelf environment or to determine how likely different elements in a design are to be noticed.
3 Fleck. M. S., Samei, E., & Mitroff, S. R. (2010). Generalised 'satisfaction of search': Adverse influences on dual-target search accuracy. *Journal of Experimental Psychology Applied*, vol. 16: pp. 60–71.
4 Tuddenham, W. J. (1962). Visual search, image organisation, and reader error in roentgen diagnosis: Studies of the psychophysiology of roentgen image perception. *Radiology*, vol. 78: pp. 694–704.

5 Levin, D. T. & Simons, D. J. (1997). Failure to detect changes to attended objects in motion pictures. *Psychonomic Bulletin Review*, vol. 4: pp. 501–506.
6 Buñuel, L. (1983). *My Last Sigh*. Alfred A. Knopf: New York.
7 Kuleshov, L. (1987). *Lev Kuleshov: Fifty Years in Films*. Moscow: Raduga. (Original work published 1920.) Lev Vladimirovich Kuleshov was a Soviet filmmaker, film theorist and founder of the Moscow Film School. He was responsible for the Kuleshov Effect, by which viewers derive more meaning from the interaction of two sequential shots than from a single shot in isolation.
8 Simons, D. J. & Levin, D. T. (2003). What makes change blindness interesting? In D. E. Irwin and B. H. Ross (eds), *The Psychology of Learning and Motivation*, vol. 42: pp. 295–322. Academic Press: San Diego, CA.
9 Simons, D. J. & Levin, D. T. (2003). What makes change blindness interesting? In D. E. Irwin and B. H. Ross (eds), *The Psychology of Learning and Motivation*, vol. 42: pp. 295–322. Academic Press: San Diego, CA.
10 Simons, D. J. (2010). Monkeying around with the gorillas in our midst: familiarity with an inattentional blindness task does not improve the detection of unexpected events. *I-Perception*, vol. 1: pp. 3–6.
11 Yao, R., Wood, K. & Simons, D. J. (2019). As if by Magic: An Abrupt Change in Motion. *Psychological Science* 30 (6).
12 Rensink, R. A. (2002). Change detection. *Annual Review Psychology*, vol. 53: pp. 245–277.
13 Drance, S. M., Berry, V. & Hughes, A. (1967). Studies of the effects of age on the central and peripheral isopters of the visual field in normal subjects. *American Journal of Ophthalmology*, vol. 63: pp. 1667–1672.
14 Ball, K. K., Beard, B. L., Roenker, D. L., et al. (1988). Age and visual search: expanding the useful field of view. *Optical Society of America* 5 (20): pp. 2210–2218.
15 Pringle, H. I., Irwin, D. E., Kramer, A. F., et al. (2001). The role of attentional breadth in perceptual change detection. *Psychonomic Bulletin and Review* 8 (1): pp. 89–95.
16 Itti, L. & Koch, C. (2000). A saliency-based mechanism for overt and covert shifts of visual attention. *Vision Research* (40): pp. 1489–1506.

17 Rothbart, M. & John, O. P. (1985). Social categorisation and behavioural episodes: A cognitive analysis of the effects of intergroup contact. *Journal of Social Issues*, vol. 41: pp. 81–104.
18 Palmer, J. (1995). Attention in visual search: Distinguishing four causes of a set size effect. *Current Directions in Psychological Science* 4 (4): pp. 118–123.
19 Mark Randon, personal interview.
20 Drew, T., Vo, M. L. H., & Wolfe, J. M. (2013). The invisible gorilla strikes again: sustained inattentional blindness in expert observers. *Psychological Science* 24: pp. 1848–1853.
21 Wolfe, J. M., Brunelli, D. N., Rubinstein, J., et al.. (2013). Prevalence effects in newly trained airport checkpoint screeners: Trained observers miss rare targets, too. *Journal of Vision* 13 (3): pp. 1–9.
22 Kelly, D. & Efthymiou, M. (2019). An analysis of human factors in fifty controlled flight into terrain aviation accidents from 2007 to 2017. *Journal of Safety Research*, vol. 69: pp. 155–165.
23 Ververs, P. M. & Wickens, C. D. (1998). Head-up displays: Effects of clutter, display intensity, and display location on pilot performance. *International Journal of Aviation Psychology*, vol. 8: pp. 377–403.
24 Kumar, J., Saini, S. S., Agrawal, D., et al. (2023). Human Factors While Using Head-Up-Display in Low Visibility Flying Conditions. *Intelligent Automation & Soft Computing* 36 (2): pp. 2411–2423.
25 Latorella, K. A. & Prabhu, P. V. (2009). A review of human error in aviation maintenance and inspection, Routledge.
26 Hamacher, B. (2022). Tragedy in the Everglades: Remembering the Crash of Eastern Airlines Flight 401, 50 Years Later. *South Florida News*.
27 Durlach, P. J. (2004). Change Blindness and Its Implications for Complex Monitoring and Control Systems Design and Operator Training. *Human–Computer Interaction*; Durlach, P. J. and Chen, J. Y. C. (2003). Visual change detection in digital military displays. Interservice/Industry Training, Simulation, and Education Conference 2003 Proceedings; Orlando, F.L., Durlach, P. J. & Meliza, L. L. (2004). The need for intelligent change alerting in complex monitoring and control systems. *Interaction between humans and autonomous systems over extended operation*. AAAI Technical Report SS-04-03. 93-97; DiVita, J.,

Obermeyer, R., Nygren, T. E., & Linville, J. M. (2004). Verification of the change blindness phenomenon while managing critical events on a combat information display. *Human Factors*, vol. 46: pp. 205–218.
28 Durlach, P. J. & Bowens, L. D. (2010). Effect of Icon Affiliation and Distance Moved on Detection of Icon Position Change on a Situation Awareness Display, *Military Psychology*, vol. 22: pp. 98–109.
29 Sarter, N. (2007). Coping with complexity through adaptive interface design. In *Human-Computer Interaction: HCI Intelligent Multimodal Interaction Environments*. Springer Berlin Heidelberg: pp. 493–498.

Chapter 3: It's Your Choice – Or Is It?

1 Johansson, P., Hall, L., Sikström, S., Tärning, B. & Lind, A. (2006). How something can be said about telling more than we can know: On choice blindness and introspection. *Consciousness and Cognition*, vol. 15: pp. 673–692. It could be argued that while important to Hall and Johansson, the choice was of little or no interest to the participants. Aware it was simply an experiment, and there was no possibility of ever meeting the person, they paid little attention when choosing. Either that or they went along with the deception out of politeness. Other psychologists have raised similar doubts about how the results of this study have been interpreted. See, for example, Sullivan-Bissett, E. and Bortolotti, L. (2019). Knowing the Unknown: Is choice blindness a case of self-ignorance? *Synthese*, vol. 198 (6): pp. 5437–5454.
2 Sagana, A., Sauerland, M. & Merckelbach, H., (2014). 'This Is the Person You Selected': Eyewitnesses' Blindness for Their Own Facial Recognition Decisions, *Applied Cognitive Psychology* 28 (5): pp. 753–764.
3 Since the early 1990s, at least in the UK, witnesses have studied the line-up through one-way mirrors to protect their identities and reduce the chances of intimidation.
4 The US Innocence Project documents mistaken identification as a factor in the wrongful conviction of 61 of the first 70 cases (87 per cent) exonerated by DNA evidence unavailable at trial. www.innocenceproject.org

5 Roughead, W. (ed.), (1924). *The Trial of Adolf Beck*. Notable British Trials. London: William Hodge; *Report of the Committee of Enquiry into the case of Mr Adolf Beck* (1904).
6 Bruce, V., Henderson, Z., Greenwood, K., Hancock, P., Burton, A. M. & Miller, P. (1999). Verification of face identities from images captured on video. *Journal of Experimental Psychology: Applied*, vol. 5: pp. 339–360.
7 Hall, L., Johansson. P., Tärning, B., Sikström, S., et al. (2010). Magic at the marketplace: Choice blindness for the taste of jam and the smell of tea. *Cognition*, doi: 10.1016/cognition.
8 Johansson, P. & Hall, L. (2008). From change blindness to choice blindness. *Psychologia*, vol. 51: pp. 142–155.
9 Hall, L., Johansson, P. & Strandberg, T. (2012). Lifting the veil of morality: Choice blindness and moral attitude reversals on a self-transforming survey. *PLOS One*, 7: e45457.
10 McLaughlin, O. & Somerville, J. (2013). Choice blindness in financial decision making. *Judgment and Decision Making* 8 (5): pp. 561–572.
11 Schanzer, J. (2013). Choice Blindness in Consumer Decision-Making. *Psychology Honors Papers* 31. digitalcommons.conncoll.edu/psychhp/31.
12 Nisbett, R. E. & Wilson, T. D. (1977). Telling more than we can know: Verbal reports on mental processes. *Psychological Review*, vol. 84: pp. 231–259.
13 Shibuya, K., Kasuga, R., Sato, N., Santa, R., Homma, C., & Miyamoto, M. (2022). Preliminary findings: Preferences of right-handed people for food images oriented to the left vs. right side. *Food Quality and Preference*, 97, 104502.
14 Jenness, A. (1932). The role of discussion in changing opinion regarding a matter of fact. *Journal of Abnormal Social Psychology*, vol. 27: pp. 279–296.
15 Asch, S. E. (1952). *Social Psychology*. New Jersey: Prentice-Hall. Reprinted 1987, Oxford University Press. Ironically, many accounts of Asch's work support assertions he intended to refute, namely the powers of independence under demanding conditions.
16 Perrin, S. & Spencer, C. (1981). The Asch effect – A child of its time. *Bulletin of the British Psychological Society*, vol. 33: pp. 405–407.
17 Krosnick, J. A., & Petty, R. E. (1995). Attitude strength: An overview. In R. E. Petty and J. A. Krosnick (eds), *Attitude*

Strength: Antecedents and Consequences (Ohio State University series on attitudes and persuasion, vol. 4): pp. 124. Hillsdale, NJ: Lawrence Erlbaum Associates.

18 Haidt, J. (2001). The Emotional Dog and Its Rational Tail: A Social Intuitionist Approach to Moral Judgment. *Psychological Review* 108 (4): pp. 814–834.

19 Horberg, E. J., Oveis, C., Keltner, D., et al. (2009). Disgust and the Moralisation of Purity. *Journal of Personality and Social Psychology* 97 (6): pp. 963–976.

20 We have encountered only one person, a dentist, who claims she could continue spitting and sipping all day. But then she spends her working life ferreting around in the mouths of others.

21 Teachman, B. A. & Saporito, J. (2009). I am going to gag: Disgust cognitions in spider and blood-injury-injection fears. *Cognitive Emotions* 23 (2): pp. 399–414.

22 Ciaramelli E. & Ghetti S. (2007). What are confabulators' memories made of? A study of subjective and objective measures of recollection in confabulation. *Neuropsychologia* 45 (7): pp. 489–500.

23 Goldenberg, G., Müllbacher, W. & Nowak, A. (1995). Imagery without perception – A case study of anosognosia for cortical blindness. *Neuropsychologia* 33 (11): pp. 1373–1382.

24 Bortolotti, L. (2018). Stranger than Fiction: Costs and Benefits of Everyday Confabulation. *Revue Philosophical Psychology* 9: pp. 227–249.

25 Pickel, K. (2004). When a lie becomes the truth: The effects of self-generated misinformation on eyewitness memory. *Memory* 12 (1): pp. 14–26.

26 Polage, D. C. (2004). Fabrication deflation? The mixed effects of lying on memory. *Applied Cognitive Psychology* 18 (4): pp. 455–465.

27 Strijbos, D. & de Bruin, L. (2015). Self-interpretation as first-person mindshaping: Implications for confabulation research. *Ethical Theory Moral Practice*, vol. 18: pp. 297–307.

28 Strandberg, T., Björklund, F., Hall, L., et al. (2019). Correction of Manipulated Responses in the Choice Blindness Paradigm: What are the Predictors? *Annual Meeting of the Cognitive Science Society*; Lachaud, L., Jacquet, B. & Baratgin, J. (2022). Reducing Choice-Blindness? An Experimental Study Comparing Experienced Meditators to Non-Meditators. *Eur. J. Investig.*

Health Psychol. Educ., vol. 12: pp. 1607–1620; Bettman, J. R., Luce, M. F. & Payne, J. W. (1998). Constructive consumer choice processes. *Journal of Consumer Research* 25 (3): pp. 187–217.

29 Schall, J. D. (2001). Neural Basis of Deciding, Choosing and Acting. *Nature Reviews Neuroscience*, vol. 2: pp. 33–41.

Chapter 4: Expectation Blind Spots

1 Triplett, N. (1900). The Psychology of Conjuring Deceptions. *The American Journal of Psychology* 11 (4): pp. 439–510.

2 Kuhn, G. & Rensink, R. (2016). The Vanishing Ball Illusion: A new perspective on the perception of dynamic events. *Cognition*, vol. 148: pp. 64–70.

3 Mehrdad Jazayeri, an MIT brain researcher, quoted in Trafton, A. (2019). How expectation influences perception: Neuroscientists find brain activity patterns that encode our beliefs and affect how we interpret the world around us. MIT News Office report, 15 July.

4 Liu, B., & Todd, J. T. (2004). Perceptual biases in the interpretation of 3D shape from shading. *Vision Research* 44 (18): pp. 2135–2145.

5 Weilnhammer, V. & Sterzer, P. (2021). Bistable perception alternates between internal and external modes of sensory processing. *Cell Press iScience*, vol. 24, 102234, 19 March.

6 Oliva, A. & Torralba, A. (2007). The role of context in object recognition. *Trends in Cognitive Science*, vol. 11: pp. 520–527.

7 Trafton, A. (2019) How expectation influences perception: Neuroscientists find brain activity patterns that encode our beliefs and affect how we interpret the world around us. MIT News Office report, 15 July.

8 McAuliff, B. D. & Bornstein, B. H. (2012). Beliefs and expectancies in legal decision making: an introduction to the Special Issue, *Psychology, Crime & Law* 18 (1).

9 James, W. (1890/1950). *The Principles of Psychology*. Vol. 1. New York: Dover.

10 Kersten, D. & Schrater, P. R. (2002). Pattern inference theory: a probabilistic approach to vision. In *Perception and the Physical World*. Edited by Mausfeld, R. & Heyer, D. John Wiley & Sons, Ltd: Chichester, pp. 191–228.

11 Feldman, J. (2015). Bayesian models of perception: a tutorial introduction. In *Oxford Handbook of Perceptual Organization*. Edited by Wagemans, J. Oxford: Oxford University Press; Witzel, C., Racey, C., & O'Regan, J. K. (2017). The most reasonable explanation of 'the dress': Implicit assumptions about illumination. *Journal of Vision*, vol. 17(2), 1: pp. 1–19.

12 Witzel, C., Racey, C., & O'Regan, J. K. (2017). The most reasonable explanation of 'the dress': Implicit assumptions about illumination. *Journal of Vision*, vol. 17(2), 1: pp. 1–19.

13 Koenderink, J. J., van Doorn, A. J., Kappers, A. M. & Todd, J. T. (2001). Ambiguity and the 'mental eye' in pictorial relief. *Perception*, vol. 30: pp. 431–448.

14 Helmholtz, H. (1867). *Handbuch der physiologischen optik*. Leipzig: Voss.

15 Nisbett, R. E. & Wilson, T. D. (1977). The halo effect: Evidence for unconscious alteration of judgments. *Journal of Personality and Social Psychology* 35 (4): pp. 250–256.

16 Thorndike, R. L. (1950). The Effect of Discussion upon the Correctness of Group Decisions, when the Factor of Majority Influence is Allowed For. *The Journal of Social Psychology* 9 (3).

17 Gilbert, D. T. & Hixon, J. G. (1991). The Trouble of Thinking Activation and Application of Stereotypic Beliefs. *Journal of Personality and Social Psychology* 60 (4): pp. 509–517.

18 Allport, G. W. (1954). *The Nature of Prejudice*. New York: Addison Wesley: pp. 20–21. Cited in Gilbert, D. T. & Hixon, J. G. (1991) op.cit.

19 McAuliff, B. D. & Bornstein, B. H. (2012). Beliefs and expectancies in legal decision making: an introduction to the Special Issue, *Psychology, Crime & Law* 18 (1).

20 Lewis, D. & Hughman, P. (1975). *Just How Just*. Secker & Warburg: London.

21 Alpert, G. P., Dunham, R. G., Stroshine. M., et al. (2006). Police Officers' Decision Making and Discretion: Forming Suspicion and Making a Stop. Report to The National Institute of Justice.

22 Spoerre, A., Moore, K. & Rice, G. E. (2021). KC police officer who fatally shot unarmed man last March will not be charged. *Kansas City Star*, 2 March.

23 Correll, J., Park, B., Judd, C. M. & Wittenbrink, B. (2002). The police officer's dilemma: Using ethnicity to disambiguate potentially threatening individuals. *Journal of Personality and*

Social Psychology 83(6): pp. 1314–1329. Acknowledging that their findings are politically controversial, Correll et al. offer these two considerations. 'Our findings should be replicated by researchers in other labs with different materials before generalisations are made. Second, our goals as psychologists include understanding, predicting, and controlling behaviour. Ultimately, efforts to control (i.e., reduce or eliminate) any ethnic bias in the decision to shoot must be based on an accurate understanding of how target ethnicity influences that decision, even if that understanding is politically or personally distasteful.' (pp. 1327–1328.)

24 Alpert, G. P., Dunham, R. G., Stroshine. M., Bennett, K. & MacDonald, J. (2006). Police Officers' Decision Making and Discretion: Forming Suspicion and Making a Stop. Report to The National Institute of Justice.

25 Correll, J., Park, B., Judd, C. M. & Wittenbrink, B. (2002). The police officer's dilemma: Using ethnicity to disambiguate potentially threatening individuals. *Journal of Personality and Social Psychology* 83(6): pp. 1314–1329.

26 Gilbert, D. T. & Hixon, J. G. (1991) The Trouble of Thinking – Activation and Application of Stereotypic Beliefs, *Journal of Personality and Social Psychology*, Vol. 60 (4): pp. 509-517.

27 Greenwald, A. G. & Krieger, L. H. (2006). Implicit bias: Scientific foundations. *California Law Review* 94 (4): pp. 945–967. You can try the test for yourself at Harvard's 'Project Implicit' website, which includes a range of fascinating IATs, shedding light on our subconscious biases. https://www.projectimplicit.net/

28 Nosek, B. A., Greenwald, A. G. & Banaji, M. R. (2005). Understanding and using the Implicit Association Test: II. Method variables and construct validity. *Personality and Social Psychology Bulletin*, vol. 31: pp. 166–180. Concerns about the reliability and interpretation of IAT results have been raised, emphasising that the test may capture associations influenced by context and cultural factors. Despite ongoing debates, the IAT remains a valuable tool for researchers examining implicit attitudes and associations, shedding light on subconscious biases that may impact behaviour and decision-making. It serves to uncover implicit associations that individuals may not readily disclose, contributing to a deeper understanding of cognitive processes and social attitudes. More recent research

has also discovered that the IAT may help uncover more in-depth personality traits, such as how likely people are to be persuaded by external information.

29 Antuano, M. J. & Mohler, S. R. (1989). Geographical disorientation: approaching and landing at the wrong airport. *Aviation, Space, and Environmental Medicine* 60 (10): pp. 996–1004.

30 Jin, L. & Lo, E. (2017). A Study of Accidents and Incidents of Landing on Wrong Runways and Wrong Airports. Report from Purdue University, West Lafayette, Indiana: pp. 107–112.

31 Andel, C., Davidow, S. L., Hollander, M. & Moreno, D. A. (2012). The economics of health care quality and medical errors. *Journal of Health Care Finance* 39 (1): pp. 39–50.

32 Institute for Safe Medication Practices (ISMP) Errors Reporting Program (MERP) (2022). The ISMP identifies anticoagulated heparin as one of the five most potentially hazardous drugs, the others being insulin, opiates and narcotics, injectable potassium chloride (or) phosphate concentrate, and sodium chloride solutions above 0.9 per cent.

33 Grissinger, M. (2012). What captures your attention? *P&T* 47 (10): pp. 542–555.

34 Nasa, P. & Majeed, N. A. (2023). Decision Fatigue among Emergency Physicians: Reality or Myth. *Indian Journal of Critical Care Medicine* 27 (9): pp. 609–610.

35 Pietzsch. M. C. & Pinquart, M. (2023). Predicting coping with expectation violations: combining the ViolEx Model and the Covariation Principle. *Frontiers in Psychology*.

36 Cumsille, P. E., Sayer, A. G. & Graham, J. W. (2000). Perceived exposure to peer and adult drinking as predictors of growth in positive alcohol expectancies during adolescence. *Journal of Consulting and Clinical Psychology*, vol. 68: pp. 531–536.

37 Colson, T. (2021). Biden said there was 'no circumstance' in which US citizens would be evacuated by helicopter five weeks before exactly that happened. Yahoo News.

38 García-Cabezas, M. Á. & Barbas, H. (2016). Anterior Cingulate Pathways May Affect Emotions Through Orbitofrontal Cortex. *Cerebral Cortex*, vol. 27: pp. 4891–4910.

39 Barbas H. (1997). Two prefrontal limbic systems: their common and unique features. In: Sakata, H., Mikami, A., & Fuster, J. M. (eds) *The Association Cortex: Structure and Function*. Amsterdam: Harwood Academic Publishing: pp. 99–115.

40 Elsey, J. W. B. & Kindt, M. (2021). Expectations of objective threats and aversive feelings in specific fears. *Nature Scientific Reports* 11:20778.
41 Rosenthal, R. & Jacobsen, L. (1968). *Pygmalion in the Classroom*. New York: Holt.
42 Kohl, H. R. (1968). *The New York Review of Books*.
43 Elashoff, J. D. & Snow, R. E. (1971). *Pygmalion in the Classroom – A Case Study in Statistical Inference: Reconsideration of the Rosenthal–Jacobson Data on Teacher Expectancy*. Worthington, OH: Charles A. Jones Publishing Company.

Chapter 5: You Could Have Fooled Me!

1 Shepard, R. N. (1990). *Mind Sights*. New York: W. H. Freeman and Co.
2 Külpe, O. (1893). *Grundriss der Psychologie: Auf experimenteller Grundlage dargestellt*. Leipzig: Wilhelm Engelmann.
3 Gregory, R. L. (1997). *Philosophical Transactions of the Royal Society of London* B 352:1121–11281997.
4 Hill, K. (2013). The Lurking Pornographer: Why Your Brain Turns Bubbles Into Nude Bodies. JREF Swift blog, 17 January.
5 Carbon, C. C. (2014). Understanding human perception by human-made illusions. *Frontiers in Human Neuroscience* 8 (566): pp. 1–6.
6 Ekroll, V., Sayim, B. & Wagemans, J. (2017). The Other Side of Magic: The Psychology of Perceiving Hidden Things. *Perspectives on Psychological Science* 12 (1): pp. 91–106.
7 Briscoe, R. E. (2011) Mental Imagery and the Varieties of Amodal Perception, *Pacific Philosophical Quarterly*, June.
8 Scherzer, T. R. & Ekroll, V. (2015). Partial modal completion under occlusion: What do modal and amodal percepts represent? *Journal of Vision* 15 (1) 22: pp. 1–20.
9 Ekroll, V., Sayim, B. & Wagemans, J. (2017). The Other Side of Magic: The Psychology of Perceiving Hidden Things. *Perspectives on Psychological Science* 12 (1): pp. 91–106.
10 Kennedy, J. M. (1988). Line endings and subjective contours. *Spatial Vision*, vol. 3: pp. 151–158.
11 Kanizsa, G. (1979). *Organization in vision: Essays on gestalt perception*. New York: Praeger.

12 Nieder, A. (2002). Seeing more than meets the eye: Processing of illusory contours in animals. *Journal of Comparative Physiology, A* 188 (4): pp. 249–260.
13 Schyns, P. & Oliva, A. (1994). From blobs to boundary edges: Evidence for time- and spatial-scale-dependent scene recognition. *Psychological Science*, vol. 5: pp. 195–200.
14 Anderson, B. L. & Winawer, J. (2005). Image Segmentation and Lightness Perception. *Nature*, vol. 434: p 80.
15 Adams, H. F. (1912). Autokinetic sensations. *Psychological Monographs*, no. 59: pp. 1–44.
16 Schweizer, G. (1857). Über das Sternschwanken. *Bulletin de la Société impériale des naturalistes de Moscou*, vol. 30: pp. 440–457.
17 Gregory, R. L. & Zangwill, O. L. (1963). The Origin of the Autokinetic Effect. *Quarterly Journal of Experimental Psychology*, vol. 15: pp. 255–261.
18 Wright, E. (2004). *Generation Kill*. New York: Transworld.
19 Aviation and Plane Crash Statistics (2020). National Transportation Safety Board.
20 Gombrich, E. H. (1959). *Art and Illusion*. Oxford: Phaidon Press.
21 Boring, E. G. (1942). *Sensation and Perception in the History of Experimental Psychology* (edited by Elliot, R. M.). Appleton-Century-Crofts Inc: New York. p. 269.
22 Hill based his cartoon on a 19th-century German postcard. A few years later, his image was used in a paper by American psychologist E. G. Boring and has remained a staple of introductory psychology courses ever since. Wright, E. (1992). The original of E. G. Boring's Young Girl/Mother-in-Law drawing and its relation to the pattern of a joke. *Perception*, vol. 21: pp. 273–275.
23 Nicholls, M. E. R., Churches, O. & Loetscher, T. (2018). Perception of an ambiguous figure is affected by own-age social biases, *Nature Scientific Reports*, vol. 8, 12661.
24 Necker, L. A. (1834) Observations on some remarkable Optical Phenomena seen in Switzerland; and on an Optical Phenomenon which occurs on viewing a figure of a Crystal or geometrical Solid. *London and Edinburgh Philosophical Magazine and Journal of Science*, Third Series.
25 Slater A. (1995). Visual perception and memory at birth. In: Rovee-Collier, C. and Lipsitt, L. P. (eds) *Advances in Infancy Research*. Ablex: Norwood, NJ, pp. 107–162.

26 Gombrich, E. H. (1959). *Art and Illusion*. Phaidon Press: Oxford.
27 In 1964, psychologist Donald Schuster was reading a flying magazine when, on an advertising page, he came across this powerful and puzzling illusion. Schuster, D. H. (1964). A new ambiguous figure: A three-stick clevis. *American Journal of Psychology* 77 (4): p. 673.
28 Gregory, R. L. (1970). *The Intelligent Eye*. New York: McGraw-Hill, p. 57.
29 Ponzo, M. (1911). Intorno ad alcune illusioni nel campo delle sensazioni tattili sull illusione di Aristotele e fenomeni analoghi. (Around some illusions in the field of tactile sensations on Aristotle's illusion and similar phenomena). *Archives Italiennes de Biologie*).
30 Shepard, R. N. (1990). *Mind Sights*. New York: W. H. Freeman and Co.
31 Stewart, D., Cudworth, C. J., & Lishman, J. R. (1993). Misperception of Time-to-Collision by Drivers in Pedestrian Accidents. *Perception* 22 (10).
32 Rose, D. & Bressan, P. (2002). Going round in circles: Shape effects in the illusion. *Spatial Vision*, vol. 15: pp. 191–203.
33 Delboeuf, J. (1865). Note sur certain illusions d'optique: Essai d'une théorie psychophysique de la manière dont l'oeil apprécie les distances and les angles (Note on certain optical illusions: Essay on a psychophysical theory concerning the way in which the eye evaluates distances and angles). *Bulletins de l'Académie Royale des Sciences Lettres et Beaux-arts de Belgique*, vol. 19: pp. 195–216.
34 Wansink, B. & Ittersum, K. (2003). Bottoms Up! The Influence of Elongation on Pouring and Consumption Volume. *Journal of Consumer Research*, vol. 30: pp. 455–463.
35 O'Shea, R. P., Chandler, N. P. & Roy, R. (2013). Dentists make larger-than-necessary holes in teeth if the teeth present a visual illusion of size. *Perception*, vol. 42: p. 16.
36 Way, L. W., Stewart, L., Gantert, W., Liu, et al. (2003). Causes and prevention of laparoscopic bile duct injuries: Analysis of 252 cases from a human factors and cognitive psychology perspective. *Annals of Surgery*, vol. 237: pp. 460–469.
37 Cornsweet, T. N. (1970). *Visual Perception*. New York: Academic Press.

38 Mach, E. (1865). On the effect of the spatial distribution of the light stimulus on the retina. In *Mach Bands: Quantitative Studies On Neural Networks in the Retina* (1965), edited by F. Ratliff (San Francisco: Holden-Day): pp. 253–271.
39 Daffner, R. H. (1989). Visual illusions in the interpretation of the radiographic image. *Current Problems in Diagnostic Radiology*, vol. 18: pp. 62–87.
40 Thomson, E. M., & Johnson, O. N. (2012). *Essentials of Dental Radiography for Dental Assistants and Hygienists*. Upper Saddle River, NJ: Pearson.
41 Lanford, L. (2014). Christ on a Cracker!: Apophenia, Pareidolia and Conspiracy Paranoia. Paper presented at Smithsonian 'The Future is Here' conference.
42 Schirber, M. (2005) Face on Mars: Why People See What's Not There, *Live Science* June 13.
43 Voss, J. L., Federmeier, K. D., & Paller, K. A. (2012). The Potato Chip Really Does Look Like Elvis! Neural Hallmarks of Conceptual Processing Associated with Finding Novel Shapes Subjectively Meaningful. *Cerebral Cortex*, vol. 2: pp. 2354–2364.
44 Mark Radon, consultant radiologist. Personal communication.
45 Frank Gaillard, founder of Radiopaedia.org. Personal communication. radiopaedia.org/articles/animal-and-animal-produce-inspired-signs
46 Alexander, R. G., Yazdanie, F., Waite, S., et al. (2021). Visual Illusions in Radiology: Untrue Perceptions in Medical Images and Their Implications for Diagnostic Accuracy. *Frontiers in Neuroscience*. Article 629469.
47 We generated these 'emerging images' algorithmically from 3D models. The algorithm scattered image elements across the image, obscuring but subtly hinting at hidden animals. Surprisingly, even advanced computer models could not predict where people would look. This shows how fast, unconscious visual processing works with slower conscious thinking. The human optical system is still much better than computers at finding meaningful things in images, even when objects are hidden.
48 Tonder, G. V. & Ejima, Y. (2000), Bottom-Up Clues in Target Finding. *Perception*, vol. 29: pp. 149–157.
49 Lanford, L. (2014). Christ on a Cracker!: Apophenia, Pareidolia and Conspiracy Paranoia. Paper presented at Smithsonian 'The Future is Here' conference.

50 By 'senses', we mean groups of sensory cells that respond to a specific physical phenomenon and correspond to particular brain regions where signals are received and interpreted.
51 Macpherson, F. (2011a). Individuating the Senses. In Macpherson (ed.) *The Senses: Classical and Contemporary Readings*, Oxford University Press.
52 Ward, J., Jonas, C., Dienes, Z., et al. (2010). Grapheme–colour synaesthesia improves detection of embedded shapes, but without pre-attentive 'pop-out' of synaesthetic colour. *Proceedings of the Royal Society B: Biological Sciences* 277 (1684): pp. 1021–1026.
53 A grapheme is a written symbol representing a sound (phoneme). This can be a single letter, such as a 't' or 'b', or a sequence of letters, such as 'ai', 'sh', 'tch'. When spoken aloud, it is a phoneme; when written, it is a grapheme.
54 See: Harrison, J. E. (2001). *Synaesthesia: The Strangest Thing.* New York: Oxford University Press; Harrison, J. E. & Baron-Cohens. (1997). Synaesthesia: a review of psychological theories. In Baron-Cohen, S. and Harrison, J. E. (eds), *Synaesthesia: Classic and Contemporary Readings*. Malden, MA: Blackwell: pp. 109–22.
55 Shams, L., Kamitani, Y. & Shimojo, S. (2000). What you see is what you hear. *Nature*, vol. 408: p. 788.
56 McGurk, H. & MacDonald, J. (1976). Hearing lips and seeing voices. *Nature* 264 (5588): pp. 746–748.
57 Massaro, D. W. & Cohen, M. M. (2000). Tests of auditory-visual integration efficiency within the framework of the fuzzy logical model of perception. *Journal of Acoustical Society of America* 108 (2): pp. 784–789.

Chapter 6: Blind Spots and Your Alien Brain

1 Hofstadter, D. (2007). *I Am A Strange Loop.* New York: Basic Books: p. 362.
2 Beaumont, W. (1838). *Experiments and Observations on the Gastric Juice and Physiology of Digestion by William Beaumont, Surgeon in the US Army.* Machlachlan & Stewart: Edinburgh.
3 Modlin, I. M. & Kidd, M. (2001). Ernest Starling and the Discovery of Secretin. *Journal of Clinical Gastroenterology* 32 (3): pp. 187–192.

4   Meissner, G. (1857). Über die Nerven der Darmwand (Via the Nerves of the Intestinal Wall). *Z Ration Med*: pp. 364–366.
5   Cryan, J. F. (2019). Address to the Annual Conference of the British Psychological Society's Psychobiology Section. *The Psychologist*, January issue.
6   Bastiaanssen, T. F. S., Cussotto, S., Claesson, M. J., et al. (2020). Gutted! Unravelling the Role of the Microbiome in Major Depressive Disorder. *Harvard Review of Psychiatry* 28 (1): pp. 26–39.
7   This is the average for adult males; females have slightly fewer at 28 trillion. Hatton, I. A. et al. (2023). The human cell count and size distribution, *PNAS* 20 (39) e2303077120.
8   Kurokawa, K. et al. (2007). Comparative metagenomics revealed commonly enriched gene sets in human gut microbiomes. *DNA Research*, vol. 14: pp. 169–181.
9   Gershon, M. D. (1998). *The Second Brain*. New York: HarperCollins.
10 Israelyan, N. et al. (2019). Effects of Serotonin and Slow-Release 5-Hydroxytryptophan on Gastrointestinal Motility in a Mouse Model of Depression. *Gastroenterology* 157 (2): pp. 507–521.
11 Camilleri, M. (2009). Serotonin in the gastrointestinal tract. *Curr Opin Endocrinol Diabetes Obesity*, vol. 16: pp. 53–59.
12 American Society of Microbiology (2023). *Gut Microbiome Communication: The Gut–Organ Axis*, 18 January.
13 See: Clarke, G., Grenham, S., Scully, P., et al. (2013). The microbiome–gut–brain axis during early life regulates the hippocampal serotonergic system in a sex-dependent manner. *Mol Psychiatry* 18 (6): pp. 666–673; Velasquez-Manoff, M. (2013). *An Epidemic of Absence: A New Way of Understanding Allergies and Autoimmune Diseases*. Scribner: New York; Benton, D., Williams, C. & Brown, A. (2007). Impact of consuming a milk drink containing a probiotic on mood and cognition. *European Journal of Clinical Nutrition*, vol. 61, 355e361; Palmer, C., Bik, E. M., DiGiulio et al. (2007). Development of the human infant intestinal microbiota. *Public Library of Science Biology* 5 (7), e177.
14 Collen, A. (2015). *10% Human: How Your Body's Microbes Hold the Key to Health and Happiness*. William Collins.
15 Caroline M. et al. (2020). Delivery Mode Affects Stability of Early Infant Gut Microbiota. *Cell Reports Medicine* 1 (9).

16 See Azad, M. B. et al. (2016). Impact of maternal intrapartum antibiotics, method of birth and breastfeeding on gut microbiota during the first year of life: a prospective cohort study. *BJOG*, vol. 123: pp. 983–993; Zeuner, B., Teze, D., Muschiol et al. (2019). Synthesis of Human Milk Oligosaccharides: Protein Engineering Strategies for Improved Enzymatic Transglycosylation. *Molecules*, vol. 24, 2033.

17 Vuong, H. E., Yano, J. M., Fung, et al. (2017). The Microbiome and Host Behaviour. *Annual Review Neuroscience* 25 (40): pp. 21–49.

18 Huang, K. D. et al. (2024). Establishment of non-Westernized gut microbiota in men who have sex with men is associated with sexual practices. *Cell Reports Medicine*, vol. 5.

19 Horwood, A. J. (ed.) (1866). *Year Books of the Reign of King Edward the First,* Vols 1–5. HMSO: London.

20 Pope, A. (1963). The Rape of the Lock, Canto 3. *The Poems* (J. Butt, ed). Yale University Press: New Haven.

21 See Piech, R. M., Pastorino, M. T. & Zald, D. H. (2010). All I saw was the cake: Hunger effects on attentional capture by visual food stimuli. *Appetite* 54 (3): pp. 579–582; Berridge, K. C. (1996). Food reward: brain substrates of wanting and liking. *Neuroscience and Biobehavioural Reviews* 20 (1): pp. 1–25.

22 Danziger, S., Levav, J. & Avnaim-Pesso, L. (2011). Extraneous factors in judicial decisions. *Proceedings of the National Academy of Sciences*, vol. 108: pp. 6889–6892.

23 The Israeli judiciary and several academic psychologists have criticised both the underlying theory and the large effect size reported in this study. 'A drop of favourable decisions from 65% in the first trial to 5% in the last trial … is equivalent to an odds ratio of 35,' comments Andreas Glöckner in his 2016 paper 'The irrational hungry judge effect revisited: Simulations reveal that the magnitude of the effect is overestimated' (*Judgment and Decision Making* 11 (6): pp. 601–610).

24 Tooley, K. L. (2020). Effects of the Human Gut Microbiota on Cognitive Performance, Brain Structure and Function: A Narrative Review. *Nutrients* 12 (10).

25 Lew, L. C., Hor, Y. Y., et al. (2018). Probiotic *Lactobacillus plantarum* P8 alleviated stress and anxiety while enhancing memory and cognition in stressed adults: A randomised, double-blind, placebo-controlled study. *Clinical Nutrition* 38 (5): pp. 2053–2064.

26 Smith, L. K. & Wissel, E. F. (2019). Microbes and the Mind: How Bacteria Shape Affect, Neurological Processes, Cognition, Social Relationships, Development, and Pathology. *Perspectives on Psychological Science* 14 (3): pp. 1–22.
27 Moloney, R. D., Desbonnet, L., Clarke, G., et al. (2013). The microbiome: stress, health and disease. *Mammalian Genome* 25 (1–2): pp. 49–74.
28 Damiani, F., Cornuti, S. & Tognini, P. (2023). The gut–brain connection: Exploring the influence of the gut microbiota on neuroplasticity and neurodevelopmental disorders. *Neuropharmacology* 231, 109491.
29 Aru, J., Suzuki, M. & Larkum, M. E. (2020), Cellular mechanisms of conscious processing. *Trends Cognitive Science*, vol. 24: pp. 814–825.
30 Wise, J. (2016). An Illusion Made FlyDubai Pilots Crash Their Plane Into the Ground. *Popular Mechanics*, April edition.
31 National Transportation Board Report to Congress (1999) p. 36.
32 Patterson, F. R. (2006). Spatial Awareness Training Systems SBIR Phase II. *Naval Air Systems Command final report*. US Government document: pp. 1–58.
33 Wise, J. (2016). An Illusion Made Fly Dubai Pilots Crash Their Plane Into the Ground. *Popular Mechanics*, April edition.
34 Patterson, F. R. (2006). Spatial Awareness Training Systems SBIR Phase II. *Naval Air Systems Command final report*. US Government document: pp. 1–58.
35 Patterson, F. R., Arnold, R. D. & Williams, H. P. (2013). Visual Perspective Illusions as Aviation Mishap Causal Factors. *17th International Symposium on Aviation Psychology*: pp. 512–517.
36 Graybiel, A. & Knepton, J. C. (1976), Sopite syndrome: A sometimes sole manifestation of motion sickness. *Aviation Space Environment Medicine*, vol. 47: p. 873. Graybiel was one of the doctors who examined John Glenn after his Mercury spacecraft circled Earth three times in February 1962 in the first orbital mission by an American. He reported that Glenn's balance scores were slightly better after the flight. He also developed a technique for operating on important cardiovascular vessels in human foetuses, laying the groundwork for repairing certain congenital heart defects. Blind spots due to sopite syndrome occur on both land and sea as well as in air flight, although usually with far less tragic consequences. In the 3rd century

BCE the Greek physician Hippocrates of Kos noted that 'sailing on the sea proves that motion disorders the body'. Indeed, the term 'nausea' – a feature of the sopite syndrome – is derived from *naus*, the Greek word for ship (e.g. nautical). Sopite syndrome victims experience decreased motor function and reduced attention span, contributing to accident-producing blind spots.

37 Patterson, F. R., Arnold, R. D. & Williams, H. P. (2013). Visual Perspective Illusions as Aviation Mishap Causal Factors. *17th International Symposium on Aviation Psychology*: pp. 512–517.

38 Koitabashi, R. & Uchida, Y. (2019). Analysing the relationship between cognition and urine storage function. *International Journal of Urological Nursing* 13 (2): pp. 51–56.

39 Moore, L. B. et al. (2017). Urinary Excretion of Sodium, Nitrogen, and Sugar Amounts Are Valid Biomarkers of Dietary Sodium, Protein, and High Sugar Intake in Nonobese Adolescents. *The Journal of Nutrition* 147 (12): pp. 2364–2373.

40 Maasalo, I., Lehtonenc, E. & Summalaa, H. (2019). Drivers with child passengers: distracted but cautious? *Accident Analysis and Prevention*, vol. 131: pp. 25–32.

41 Tuk, M. A., Trampe, D. & Warlop, L. (2004). Inhibitory Spillover: Increased Urination Urgency Facilitates Impulse Control in Unrelated Domains. *Psychological Science* 22 (5): pp. 627–633.

42 Van den Bergh, B., Schmitt, J., Dewitte, S. & Warlop, L. (2009). Bending Arms, Bending Discounting Functions – How Motor Actions Affect Intertemporal Decision-Making. Report from Erasmus University, Rotterdam School of Management.

43 Friedman, R. S. & Förster, J. (2000). The Effects of Approach and Avoidance Motor Actions on the Elements of Creative Insight. *Journal of Personality and Social Psychology* 79 (4): pp. 477–492.

44 Li, X. (2008). The Effects of Appetitive Stimuli on out-of-Domain Consumption Impatience. *Journal of Consumer Research* 34 (5): pp. 649–656.

45 Barsalou, L. W. (2008). Grounded Cognition. *Annual Review of Psychology*, vol. 59: pp. 617–45.

46 Förster, J., Friedman, R. S., Özelsel, A. & Denzler, M. (2006). Enactment of Approach and Avoidance Behavior Influences

the Scope of Perceptual and Conceptual Attention. *Journal of Experimental Social Psychology* 42 (2): pp. 133–146.
47 Loewenstein, G. (1996). Out of control: Visceral influences on behaviour. *Organisational Behaviour and Human Decision Processes*, vol. 65: pp. 272–292.
48 Ariely, D. & Loewenstein, G. (2006). The heat of the moment: The effect of sexual arousal on sexual decision making. *Journal of Behavioural Decision Making*, vol. 19: pp. 87–98.
49 Tuk, M. A., Trampe, D. & Warlop, L. (2004). Inhibitory Spillover: Increased Urination Urgency Facilitates Impulse Control in Unrelated Domains. *Psychological Science* 22 (5): pp. 627–633.

Chapter 7: Do You See What I See?

1 Noë, A. (ed.) (2002). *Is the Visual World a Grand Illusion?* Imprint Academic.
2 Wolfe, J. M. (1997). Visual experience: Less than you think, more than you remember. In: *Neuronal basis and psychological aspects of consciousness*, ed. C. Taddei-Ferretti. World Scientific; (1999) Inattentional amnesia. In: *Fleeting Memories*, ed. V. Coltheart. MIT Press.
3 Boudri, J. C. (1975). Maurpertuis et la critique de la métaphysique. *Actes des journée Maurpertuis*. Paris: Virin: pp. 79–9. Maurpertuis's claim to fame is that he helped determine the world's shape. Up until his time, it was widely believed to be a perfect sphere. He mathematically showed that it resembles a slightly squashed orange, flatter at the poles and broader at the equator.
4 Bostrom, N. (2003). Are you living in a computer simulation? *Philosophical Quarterly* 53 (211): pp. 243–255. Oxford University philosopher Nick Bostrom suggests that 'the vast majority of minds like ours do not belong to the original race but to people simulated by the advanced descendants of an original race'. If this were the case, we would be rational in thinking that we are likely among the simulated minds rather than among the original biological ones.
5 Hürter. T. (2022). *The Age of Uncertainty – How Physics Changed the Way We See the World 1895–1945*. London: Scribe.
6 Durga, S. (2023). Experience Cow Vision. SR Publications: www.srpublication.com/experience-cow-vision.

7   Our eyes can perceive 60 flashes of light per second, known as the 'flicker rate'. With a flicker rate of 250, a fly sees four times faster than we do.
8   Finlay, B. L., Clancy, B. & Kingsbury, M. A. (2003). The developmental neurobiology of early vision. In: Hopkins, B., and Johnson, S. P. (eds) *Neurobiology of Infant Vision*. Praeger Publishers: Westport. pp. 1–41.
9   Stein, B. E., Perrault, T. J., Jr, Stanford, T. R. et al. (2009). Postnatal Experiences Influence How the Brain Integrates Information from Different Senses. *Frontiers in Integrative Neuroscience* 3 (21). Johnson, S. P. (2001). Neurophysiological and psychophysical approaches to visual development. In: Kalverboer, A. F, and Gramsbergen, A. (eds) *Handbook of Brain and Behaviour in Human Development*. Elsevier: Amsterdam. pp. 653–675; Prechtl, H. F. R. (2001). Prenatal and early postnatal development of human motor behaviour. In: Kalverboer, A. F. and Gramsbergen, A. (eds) *Handbook of Brain and Behaviour in Human Development*. Elsevier: Amsterdam, pp. 415–427; Haith, M. M. (1980). *Rules that Babies Look By: The Organization of Newborn Visual Activity*. Hillsdale, NJ: Erlbaum.
10 Okawa, H., & Sampath, A. P. (2007). Optimisation of single-photon response transmission at the rod-to-rod bipolar synapse. *Physiology*, vol. 22: pp. 279–286.
11 Livingstone, M. (2003). Light Vision, *Harvard Medical Alumni Bulletin*: p. 20.
12 Salthouse, T. A. & Ellis, C. L. (1980) Determinants of Eye-Fixation Duration, *The American Journal of Psychology* Vol. 93 (2): pp. 207-234.
13 Hahnel, C., Goldhammer, F., Naumann, J. et al. (2016). Effects of linear reading, basic computer skills, evaluating online information, and navigation on reading digital text. *Computers in Human Behaviour*, vol. 55, part A: pp. 486–500.
14 Rayner, K. & Pollatsek, A. (1989). *The Psychology of Reading*. Englewood Cliffs, NJ: Prentice Hall.
15 Weger, U. W. & Inhoff, A. W. (2006). Attention and eye movements in reading: inhibition of return predicts the size of regressive saccades. *Psychological Science* 17 (3): pp. 187–191.
16 Fischer, M. H. (1999). An Investigation of Attention Allocation During Sequential Eye Movement Tasks. *The Quarterly Journal of Experimental Psychology* Section A 52 (3): pp. 649–677.

17 Wood, E. A. (2000). Working in the Fantasy Factory: The Attention Hypothesis and the Enacting of Masculine Power in Strip Clubs. *Journal of Contemporary Ethnography* 29 (1): pp. 5–31.

18 Tuomisto, T., Hetherington, M. M., Morris, M. F., et al. (1999). Psychological and physiological characteristics of sweet food 'addiction'. *International Journal of Eating Disorders*, vol. 25: pp. 169–175.

19 Hetherington, M. M. & MacDiarmid, J. I. (1993). Chocolate addiction: a preliminary study of its description and its relationship to problem eating. *Appetite*, vol. 21: pp. 233–246.

20 Fugère, M. A., Ciccarelli, N. C., & Cousins, A. J. (2023). The importance of physical attractiveness and ambition/intelligence to the mate choices of women and their parents. *Evolutionary Behavioral Sciences*.

21 Barber, N. (2018). Why Looks Still Matter as Women Gain Power: Physical attractiveness is affected by evolutionary aesthetics. *Psychology Today*.

22 Edison Studios made the earliest Westerns, a series of short, single-reel silents, in 1894. In 1903, Edwin S. Porter directed *The Great Train Robbery*, generally considered the start of the Western genre.

23 Purves, D., Paydarfar, J. A., & Andrews, T. J. (1996). The wagon wheel illusion in movies and reality. *Proceedings of the National Academy of Science* 93 (8): pp. 3693–3697.

24 See: VanRullen, R., Reddy, L, & Koch, C. (2006). The continuous wagon wheel illusion is associated with changes in electroencephalogram power at approximately 13 Hz. *Journal of Neuroscience* 26 (2): pp. 502–507; Busch, N. A., Dubois, J. & VanRullen, R. (2009). The Phase of Ongoing EEG Oscillations Predicts Visual Perception. *The Journal of Neuroscience* 29 (24): pp. 7869–7876; Alamia, A. & VanRullen R. (2019). Alpha oscillations and travelling waves: Signatures of predictive coding? *PLoS Biology* 17 (10): e3000487.

25 Lindberg, D. C. (1976). *Theories of Vision from Al-Kindi to Kepler.* University of Chicago Press: p. 2003.

26 Noë, A. & O'Regan, K. (2000). Perception, Attention and the Grand Illusion. *Psyche* 6 (15).

27 Descartes, R. (1637/1902). La Dioptrique. In: *Oeuvres de Descartes*, vol. 6. Edited by Adam, C. & Tannery, P. Léopold Cerf: Paris, p. 101.

28 Lettvin, J. Y., Maturana, H. R., McCulloch, W. S. et al. (1959). What the frog's eye tells the frog's brain. *Proceedings of the IRE*. The simplicity of the frog's visual system comes at a price. They can only detect something that moves, and will starve to death if surrounded by food that remains stationary. Having no fovea, they cannot see anything in sharp focus and only remember an object that remains in sight. When threatened, their sole survival strategy is leaping into anywhere darker.

29 O'Regan, J. K. & Noë, A. (2001). A sensorimotor account of vision and visual consciousness. *Behavioural and Brain Sciences*, vol. 24: pp. 939–1031.

30 O'Regan, J. K. & Noë, A. (2001). A sensorimotor account of vision and visual consciousness. *Behavioural and Brain Sciences*, vol. 24: pp. 939–1031.

31 Rodriguez, E., George, N., Lachaux, J. P. et al. (1999) Perception's shadow: long-distance synchronisation of human brain activity. *Nature* 397: pp. 430–433.

32 Blom, T., Feuerriegel, D., Johnson, P., et al. (2023). Predictions drive neural representations of visual events ahead of incoming sensory information. *PNAS*.

33 Hawkins, J. & Blakeslee, S. (2004). *On Intelligence*. Owl Books/Times Books: p. 158.

34 Possibly as a result of his family's treatment by the communists, Jakob von Uexküll went on to develop his *Umwelt* by becoming a Nazi and a strong antisemite. He died in 1944.

35 Quoted in Agamben, G. (2004). *The Open: Man and Animals*. Stanford University Press: p. 46. It is not that the tick likes the taste of blood; as von Uexküll showed in his laboratory, it has no taste buds and is attracted solely by the fluid's 37°C warmth – it will drink anything at that temperature!

36 Nagel, T. (1974). What Is It Like to Be a Bat? *The Philosophical Review* 83 (4): pp. 435–450.

37 Witzel, C., Racey, C. & O'Regan, J. K. (2017). The most reasonable explanation of 'the dress': Implicit assumptions about illumination. *Journal of Vision* 17 (2) 1: pp. 1–19.

Chapter 8: Why We Never Forget a Face

1 Bruce, V. & Young, A. W. (2012). *Face Perception*. Hove: Psychology Press.

2 Bruce, V. & Young, A. W. (2012). *Face Perception*. Hove: Psychology Press.
3 Keil, M. S. (2009). 'I Look in Your Eyes, Honey': Internal Face Features Induce Spatial Frequency Preference for Human Face Processing. *PLoS Computational Biology* 5 (3).
4 Reid, V. M., Dunn, K., Young, R. J., et al. (2017). The Human Foetus Preferentially Engages with Face-like Visual Stimuli. *Current Biology*, vol. 27: pp. 1825–1828.
5 Kobylkov, D., Zanon, M. & Vallortigara, G. (2024). Innate face-selectivity in the brain of young domestic chicks. *Proceedings of the National Academy of Sciences* 21 (40), e2410404121.
6 Sergent, J. (1984). An investigation into component and configural processes underlying face perception. *British Journal of Psychology*, vol. 75: pp. 221–242.
7 Richler, J. J., Mack, M. L., Gauthier, I., et al. (2009). Holistic processing of faces happens at a glance. *Vision Research* 49 (23): pp. 2856–2861.
8 Burton, A. M., Schweinberger, S. R., Jenkins, R. et al. (2015). Arguments against a 'configural processing' account of familiar face recognition. *Perspectives on Psychological Science* 10 (4): pp. 482–96.
9 Tanaka, J. W., & Simonyi, D. (2016). The 'parts and wholes' of face recognition: A review of the literature. *Quarterly Journal of Experimental Psychology* 69 (10): pp. 1–16.
10 Sergent, J. (1984) An investigation into component and configural processes underlying face perception. *British Journal of Psychology*, vol. 75: pp. 221–242.
11 Sinha, P., Balas, B. & Ostrovsky, Y. (2007). Discovering faces in infancy. *Journal of Vision* 7 (9): pp. 569–569a.
12 Ekman, P. (1993). Facial expression and emotion. *American Psychologist*, vol. 48: pp. 384–392.
13 Matsumoto. D. & Willingham, B. (2009). Spontaneous Facial Expressions of Emotion of Congenitally and Noncongenitally Blind Individuals. *Journal of Personality and Social Psychology* 96 (1): pp. 1–10.
14 Lewis, D. (2015). *The Secret Language of Your Child*. London: Souvenir Press.
15 Thomson, P. (1980). Margaret Thatcher: a new illusion. *Perception*, vol. 9: pp. 483–484.

16 Mooney, C. M. (1957). Age in the development of closure ability in children. *Canadian Journal of Psychology/Revue canadienne de psychologie* 11 (4): pp. 219; Carbon, C.-C. Grüter, M. & Grüter, T. (2013). Age-dependent face detection and face categorisation performance. *PLOS One* 8 (10): e79164, 2013.

17 Schwiedrzik, C. M., Melloni, L. & Schurger, A. (2018). Mooney face stimuli for visual perception research. *PLOS One* 13 (7): e0200106.

18 Baldwin, M. W., Carrell, S. E. & Lopez, D. F. (1990). Priming Relationship Schemas: My Advisor And The Pope Are Watching Me From The Back Of My Mind. *Journal of Experimental Social Psychology*, vol. 26: pp. 435–454.

19 Parvizi, J., Jacques, C., Foster, B. L, et al. (2012). Electrical Stimulation of Human Fusiform Face-Selective Regions Distorts Face Perception. *Journal of Neuroscience* 32 (43): pp. 14915-14920.

20 Kanwisher, N., McDermott, J., & Chun, M. M. (1997). The fusiform face area: a module in human extrastriate cortex specialised for face perception. *Journal of Neuroscience*, vol. 17: pp. 4302–4311.

21 The term first appeared in David Inglis's 1899 paper Moral Imbecility, published in *Transactions of the Michigan State Medical Society* 23: 377–387.

22 Laurence, S., Eyre, J. & Strathie, A. (2021). Recognising Familiar Faces Out of Context. *Perception* 50 (2): pp. 174–177.

23 Thomson, D. M. (1986). Face recognition: More than a feeling of familiarity? In H. D. Ellis, M. A. Jeeves, F. Newcombe, & A. Young (eds), *Aspects of face processing*. Martinus Nijhoff: pp. 118–122.

24 Young, A. W., Hay, D. C., & Ellis, A. W. (1985). The faces that launched a thousand slips: Everyday difficulties and errors in recognizing people. *British Journal of Psychology*, Vol. 76: pp. 495–523.

25 Clifford, B. R. & Hollin, C. R. (1981). Effects of the type of incident and the number of perpetrators on eyewitness memory. *Journal of Applied Psychology*, vol. 66: pp. 364–370.

26 Davidson, P. S. R., Cook, S. P., Glisky, E. L., et al. (2005). Source Memory in the Real World: A Neuropsychological Study of Flashbulb Memory. *Journal of Clinical and Experimental Neuropsychology* 27 (7): pp. 915–929.

27 Zsok, F., Haucke, M., De Wit, C. Y. et al. (2017). What kind of love is love at first sight? An empirical investigation. *Personal Relationships*, vol. 24: pp. 869–885.
28 Office for National Statistics UK Divorce Rate, Census 2021.
29 US Office for Divorce Statistics 2022.
30 Zsok, F., Haucke, M., de Wit, C. et al. (2017) What kind of love is love at first sight? An empirical investigation. *Personal Relationships*, vol. 24: pp. 869–885.
31 Fisher, H. E. (1992). *Anatomy of Love: The Natural History of Monogamy, Adultery, and Divorce*. New York, NY: Norton.
32 Brand, S., Luethi, M., von Planta, A., et al. (2007). Romantic love, hypomania, and sleep pattern in adolescents. *Journal of Adolescent Health*, vol. 41: pp. 69–76.
33 Jacques-Tiura, A. J., Abbey, A., Parkhill, M. R., et al. (2007). Why Do Some Men Misperceive Women's Sexual Intentions More Frequently Than Others Do? An Application of the Confluence Model. *Personality & Social Psychology Bulletin* 33 (11): pp. 1467.
34 Lewis, D. (2017) The Secrets of Love at First Sight. Mindlab study conducted on behalf of a commercial client.
35 Darwin, C. & Wallace A. (1858) On the Tendency of Species to Form Varieties; and On the Perpetuation of Varieties and Species by Natural Means of Selection. Communicated by: Charles Lyell and J. D. Hooker, to the Linnean Society on 30 June.
36 See: Norgan, N. (1997). The beneficial effects of body fat and adipose tissue in humans. *International Journal of Obesity*, Vol. 21(9): pp. 738-746 and Lassek, W. D., & Gaulin, S. J. (2008). Waist-hip ratio and cognitive ability: Is gluteofemoral fat a privileged store of neurodevelopmental resources? *Evolution and Human Behaviour*, Vol. 29(1): pp. 26-34.
37 Buss, D. M. (1994) Strategies of Human Mating, *American Scientist* Vol. 15 (2): pp. 238 – 249.
38 Bateman, A. J. (1948). Intra-sexual selection in Drosophila. *Heredity* Vol. 2 (3): pp. 349-368.
39 Fromhage, L. &. Jennions, M. D. (2016) Coevolution of parental investment and sexually selected traits drives sex-role divergence, *Nature Communications* 7:12517.
40 See: Cronin, H. (1992) *The Ant and the Peacock: Altruism and Sexual Selection From Darwin to Today.* Cambridge University Press: Cambridge; Buss, D. M. & Schmidt, D. (1993) Sexual

strategies theory: an evolutionary perspective on human mating, *Psychological Review* Vol. 100 (2): pp. 204-232.
41 Wiederman, M. W. (1993). Evolved gender differences in mate preferences: Evidence from personal advertisements. *Ethology and Sociobiology*, Vol. 14(5): pp. 331–351.
42 See: Asendorpf, J. B., Penke, L., & Back, M. D. (2011). From Dating to Mating and Relating: Predictors of Initial and Long–Term Outcomes of Speed–Dating in a Community Sample. *European Journal of Personality*. Little, A. C. (2015). Attraction and human mating. In Sangrador, P., and Yela, C. (2000). 'What is beautiful is loved': Physical attractiveness in love relationships in a representative sample. *Social Behaviour and Personality*, Vol. 28: pp. 207–218.
43 Sangrador, P., & Yela, C. (2000). 'What is beautiful is loved': Physical attractiveness in love relationships in a representative sample. *Social Behaviour and Personality*, Vol. 28: pp. 207–218.
44 Coetzee, V., Greeff, J. M., Stephen, I. D. et al. (2014). Cross-Cultural Agreement in Facial Attractiveness Preferences: The Role of Ethnicity and Gender. *PLOS One* 9 (7): e99629.
45 Wolf, N. (1991). *The Beauty Myth*. New York: Morrow.
46 Lewis, D. & Hughman, P. (1974) *Just How Just?* London: Secker & Warburg.
47 Langlois, J. H. & Roggman, L. A. (1990). Attractive faces are only average. *Psychological Science*, vol. 1: pp. 115–121.
48 Penton-Voak, I. S., Perrett, D. I. & Peirce, J. W. (1999). Computer graphic studies of the role of facial similarity in judgements of attractiveness. *Current Psychology*, vol. 18: pp. 104–117.
49 Conroy-Beam, D. & Buss, D. M. (2017). Euclidean distances discriminatively predict short-term and long-term attraction to potential mates. *Evolution and Human Behaviour* 38 (4): pp. 442–450.
50 Maurer, D., Le Grand, R. & Mondloch, C. J. (2002). The many faces of configural processing. *Trends in Cognitive Sciences* 6 (6): pp. 255–260.
51 Hume, D. K., & Montgomerie, R. (2001). Facial attractiveness signals different aspects of 'quality' in women and men. *Evolution and Human Behaviour*, vol. 22: pp. 93–112.
52 Fink, B., Neave, N., Manning, J. T. et al. K. (2006). Facial symmetry and judgements of attractiveness, health and personality. *Personality and Individual Differences* 41 (3): pp. 491–499.

53 Cañigueral, R. & Hamilton, A. F. C. (2019). The Role of Eye Gaze During Natural Social Interactions in Typical and Autistic People. *Frontiers in Psychology* 10 (560).
54 McDonald, J. H. (2011). Myths of Human Genetics. University of Delaware. udel.edu/~mcdonald/mytheyecolor.html
55 Laeng, B., Mathisen, R. S & Johnsen, J. A. K. (2007). Why do blue-eyed men prefer women with the same eye colour? *Behavioural Ecology Sociobiology*, vol. 61: pp. 371–384.
56 Blue, grey, green and hazel eyes are typically found in those of European ancestry; other people's eyes are various shades of brown. Caucasian infants are typically born with clear blue eyes. As their eyes are exposed to sunlight, melanin pigment gradually changes the iris to its adult colouring, a process completed by age three.
57 Davenport, G. C. & Davenport, C. B. (1907). Heredity of eye colour in man. *Science*, vol. 26: pp. 589–592.

Chapter 9: Buying Blind Spots

1 Unpublished Mindlab International study conducted between 2020 and 2023.
2 Rodgers, E. (2023). Grocery Store Statistics: Where, When, & How Much People Grocery Shop. Drive Research blog.
3 Crockett, Z. (2024). The customers who repeatedly buy doomed products. *The Hustle*.
4 Lewis, D. (2013). *Brain Sell*. London: Nicholas Brealey Publishing.
5 Schwartz, B. (2013). *The Paradox of Choice: Why Less is More*. Harper Perennial.
6 Iyengar, S. S. & Lepper, M. (2000). When Choice is Demotivating: Can One Desire Too Much of a Good Thing? *Journal of Personality and Social Psychology* 79 (6): pp. 995–1006.
7 Interview with Kate Nightingale, by Duncan Smith on Mindlab Intelligent Insights, 2023.
8 Rodgers, E. (2023) Grocery Store Statistics: Where, When, & How Much People Grocery Shop. Drive Research blog.
9 Anonuevo, K. (2022). Will You Spend Nearly a Year of Your Life Grocery Shopping? *Verge*.
10 Alter, A. L., Oppenheimer, D. M. & Epley, N. (2007). Overcoming Intuition: Metacognitive Difficulty Activates

Analytic Reasoning. *Journal of Experimental Psychology: General* 136 (4): pp. 569–576.
11. Whan Park, C., Iyer, E. S. & Smith, D. C. (1989). The Effects of Situational Factors on In-Store Grocery Shopping Behaviour: The Role of Store Environment and Time Available for Shopping. *Journal of Consumer Research*, vol. 15: pp. 422–433.
12. Wyer, R. S. & Hartwick, J. (1980). The Role of Information Retrieval and Conditional Inference Processes in Belief Formation and Change. In: *Advances in Experimental Social Psychology*, vol. 13 (ed. Leonard Berkowitz). New York: Academic Press.
13. Commerce Statistics For 2024. *Forbes Magazine*.
14. Malone, C. (2023). Combatting Online Payment Fraud. Juniper Research.
15. Based on interviews with over 1,000 US respondents aged 18–65+ by Splitit and Google Consumer.
16. Larson, J. S., Bradlow, E. T. & Fader, P. S. (2005). An Exploratory Look at Supermarket Shopping Paths. *International Journal of Research in Marketing* 22 (4): pp. 395–414.
17. Sorensen, H. (2003). The Science of Shopping. *Marketing Research* 15 (3): pp. 30–35.
18. Sorensen, H. (2009). *Inside the Mind of the Shopper*. New Jersey: Wharton School Publishing.
19. Corstjens, J. & Corstjens, M. (2004). *The Battle for Mindspace and Shelfspace*. Gordon Wiley: Chichester.
20. Packard, V. (1958). *The Hidden Persuaders*. Penguin: London.
21. (1957) Diddling the Subconscious, *Nation*, 5 October: pp. 206–207.
22. Cousins, N. (1957) Smudging the Subconscious, *Saturday Review*, 5 October: pp. 20–40.
23. Crandall, K. B. (2006). Invisible commercials and hidden persuaders: James M. Vicary and the subliminal advertising controversy of 1957. Undergraduate Honors Thesis, University of Florida.
24. Garfield, B. (2000). Subliminal seduction and other urban myths. *Advertising Age*, 18 September.
25. Appel, M. (2017). Of sex and ice cubes: The great subliminal advertising scare. WARC: https://tinyurl.com/43ke64pr
26. Era, V., Candidi, M. & Aglioti, S. M. (2015). Subliminal presentation of emotionally negative vs positive primes

increases the perceived beauty of target stimuli. *Experimental Brain Research*, vol. 233: pp. 3271–3281.
27 Wang, Y. & Zhang, Q. (2016). Affective Priming by Simple Geometric Shapes: Evidence from Event-related Brain Potentials. *Frontiers in Psychology*, vol. 7.
28 Diedrick, B. (2024). *Top 10 Product Placements in Movie History.* Ampersand.
29 Dijksterhuis, A., Aarts, H. & Smith, P. K. (2004). The Power of the Subliminal: Subliminal Persuasion and Other Potential Applications. In *The New Unconscious*, Ran R. Hassin, et al. (eds), Oxford University Press.
30 White, M. (1997). Toy Rover Sales Soar Into Orbit: Mars Landing Puts Gold Shine Back Into Space Items. *Arizona Republic*, 12 July: E1.
31 Berger, J. and Fitzsimons, G. (2008). Dogs on the Street, Pumas on Your Feet: How Cues in the Environment Influence Product Evaluation and Choice. *Journal of Marketing Research* 45 (1): pp. 1–14.
32 Berger, J. & Fitzsimons, G. (2008). Dogs on the Street, Pumas on Your Feet: How Cues in the Environment Influence Product Evaluation and Choice. *Journal of Marketing Research* 45 (1): pp. 1–14.
33 Carter, T. J., Ferguson, M. J. & Hassin, R. R. (2008). A Single Exposure to the American Flag Shifts Support Toward Republicanism up to 8 Months Later. *Psychological Science* 22 (8): pp. 1011–1018.
34 Meyer, D. E., & Schvaneveldt (1971) Facilitation in recognizing pairs of words: Evidence of a dependence between retrieval operations, *Journal of Experimental Psychology*, Vol. 90 (2): pp. 227 -234.
35 Hutchison, K. A. & Balota, D. A. et al (2013) The semantic priming project, *Behavioural Research*, Vol. 45: pp. 1099–1114.
36 Papies, E. K. & Hamstra, P. (2010). Goal Priming and Eating Behaviour: Enhancing Self-Regulation by Environmental Cues. *Health Psychology* 29 (4): pp. 384–388.
37 Bahrami, B., Lavie, N. & Rees, G. (2007). Attentional Load Modulates Responses of Human Primary Visual Cortex to Invisible Stimuli. *Current Biology*, vol. 17: pp. 509–513.
38 Miller, K. (2023) There, at the back of the classroom, sat a person wearing a black fabric bag from the top of his head down to his bare ankles and feet. *Oregon Starter Magazine*, Winter issue.

39 Zajonc, R. B., (1968). Attitudinal effects of mere exposure. *Journal of Personality and Social Psychology Monographs* 9 (2, 2): pp. 1–27.
40 Stafford, T., & Grimes, A. (2012). Memory Enhances the Mere Exposure Effect. *Psychology & Marketing* 29 (12): pp. 995–1003.
41 Ferraro, R., Bettman, J. R. & Chartrand, T. L. (2008). The power of strangers: The effect of incidental consumer brand encounters on brand choice. *Journal of Consumer Research*. vol. 35: pp. 729–741.
42 Malmgren, P. (2015). *Signals: How Everyday Signs Can Help Us Navigate the World's Turbulent Economy*. Weidenfeld & Nicholson.
43 Liberto, D. (2023). Shrinkflation: What It Is, Reasons for It, How to Spot It. www.investopedia.com/terms/s/shrinkflation.asp
44 Ellmore, A. (2023). Haven't I Seen You Somewhere Before? A Look at Supermarket Copycat Branding. Dawn Ellmore Employment.

Chapter 10: Business Blind Spots

1 Estrin, J. (2015). Kodak's First Digital Moment. *New York Times,* 12 August.
2 Estrin, J. (2015). Kodak's First Digital Moment. *New York Times,* 12 August.
3 Summerfield, R. (2013). Eastman Kodak finally exits bankruptcy. *Financier Worldwide*.
4 Edkins, G. (2016). Human Factors, Human Error, and the Role of Bad Luck in Incident Investigations. *Safety Wise*.
5 You can find further blind spot-induced marketing failures on our blog at themindlab.co.uk/blog/museum-of-failed-products
6 Berlin, I. (1953). *The Hedgehog and the Fox: An essay on Tolstoy's view of history*. Weidenfeld & Nicolson: London.
7 Borum, R. (2014). The Fox and the Hedgehog: Contrasting Approaches to Anticipating the Environment. *Infinity Journal* 3 (4): pp. 30–33.
8 Tetlock, P. (2005). *Expert Political Judgement: How Good Is It? How Can We Know?* Princeton University Press.
9 Allport, G. W. (1954). *The Nature of Prejudice*. New York: Addison Wesley: pp. 20–21.
10 Tetlock, P. (2005). *Expert Political Judgement: How Good Is It? How Can We Know?* Princeton University Press.

11 Osteryoung, J. (2012). Business problems you ignore won't solve themselves. *Florida Small Business*.

12 Wickelgren, W. (1974). *How To Solve Problems: Elements of a Theory of Problems and Problem Solving*. W. H. Freeman & Co: San Francisco.

13 Propp, K. M. (1995). An experimental examination of biological sex as a status cue in decision-making groups and its influence on information use. *Small Group Research* 26 (4): pp. 451–474.

14 Kobvich, Y. (2023). The Women Soldiers Who Warned of a Pending Hamas Attack – and Were Ignored. *Haaretz*, 20 November.

15 Duncker, K. (1945). On Problem Solving. *Psychological Monographs*, vol. 58: pp. 1–113.

16 Jarwan, F. (2002). *Teaching thinking: Concepts and applications* (1st Ed.). Al Ain: University Book House.

17 Al-Khatib, B. A. (2012). The Effect of Using Brainstorming Strategy in Developing Creative Problem Solving Skills among Female Students in Princess Alia University College, *American International Journal of Contemporary Research* 2 (10): pp. 29–38.

18 Inspired by Brian Brushwood and Wayne Hoffmann from Scam Nation: www.youtube.com/watch?v=QKTagrpNNt8

19 Carr, N. (2008). How many computers does the world need? Fewer than you think. *Guardian*.

20 Buntz, B. (2024). Pushing the frontier of drug discovery with the world's most powerful supercomputer. *Drug Discovery & Development*, 21 February.

21 Chen, S. (2024). Exascale computers: 10 Breakthrough Technologies 2024. *MIT Technology Review*.

22 Hamilton, H. (2017). If cars developed at the same pace as computers, they'd be this fast. *Electronic Products*, 10 February.

23 SG Analytics (2020); Desjardins, J. (2019). How much data is generated each day? *World Economic Forum* and *Northeastern University Graduate Programs*, 17 April; Finances Online (2022). 53 Important Statistics About How Much Data Is Created Every Day.

24 Calder, S. & Stevenson, C. (2023). Thousands of passengers face fresh flight cancellations in aftermath of global IT outage. *Independent*, 20 July.

25 Prahl, A. & Van Swol, L. M. (2016). The Computer Said I Should: How Does Receiving Advice From a Computer Differ From Receiving Advice From a Human. Paper presented at the 66th Annual International Communication Association Conference, Fukuoka, Japan.

26 Shapiro, S. J. (2023). *Fancy Bear Goes Phishing: The Dark History of the Information Age in Five Extraordinary Hacks.* Penguin Press.

27 Antonio Garcia Martinez, A. G. (2017). *Chaos Monkey: Obscene Fortune and Random Failure in Silicon Valley.* HarperCollins.

28 Iriondo, R. (2018). Amazon Scraps Secret AI Recruiting Engine that Showed Biases Against Women. Carnegie Mellon University School of Computer Science report. https://tinyurl.com/mvc594da

29 Lazer, D., Kennedy, R., King, G. et al. (2014). The Parable of Google Flu: Traps in Big Data Analysis. *Science*, vol. 343: pp. 1203–1205: Kandula, S. and Shaman, J. (2019). Reappraising the utility of Google Flu Trends. *PLoS Computational Biology* 15 (8), 2 August: e1007258.

30 (2018). Cambridge Analytica and Facebook: The Scandal so Far, *New York Times,* 4 April. https://tinyurl.com/4rkwm58b

31 Blouin, L. (2023). AI's mysterious 'black box' problem, explained. Press release from University of Michigan-Dearborn, 6 March. umdearborn.edu/news/ais-mysterious-black-box-problem-explained

32 Robinson, S. (2024). What are data silos and what problems do they cause? TechTarget. www.techtarget.com/searchdatamanagement/definition/data-silo

33 Newcomb, T. M. (1956). The prediction of interpersonal attraction. *American Psychologist, vol. 11* (11): pp. 575–586. doi.org/10.1037/h0046141

34 Although federal election receipts show that Bankman-Fried donated almost exclusively to Democrats, he claimed on a November phone call with YouTuber Tiffany Fong that he donated an equal amount to Republicans and Democrats. https://tinyurl.com/3cfbcmwb

35 FTX crypto boss Sam Bankman-Fried denied bail in Bahamas, BBC report December 2022, https://tinyurl.com/2bxn3xd8

36 Warren, R., Price, J., VanDerWal, J., Cornelius, S., & Sohl, H. (2018) *The implications of the United Nations Paris Agreement on Climate Change for Globally Significant Biodiversity Areas, Climatic Change.*

Chapter 11: Magic Blind Spots

1 Israeli-British Uri Geller gained international fame after claiming to possess psychic powers instead of following standard stage-magician practice.
2 Richardson, B. (2023). Magic and Mentalism of Barrie Richardson. Alakazam Magic, alakazam.co.uk/zh/products/magic-and-mentalism-of-barrie-richardson-1-by-barrie-richardson-and-ll-video-download
3 Freudenburg, W. R. & Alario, M. (2007). Weapons of mass distraction: Magicianship, misdirection, and the dark side of legitimation. *Sociological Forum* 22 (2): pp. 146–173.
4 Pailhès, A., Rensink, R. A. & Kuhn, G. (2020). A Psychologically Based Taxonomy of Magicians' Forcing Techniques: How magicians influence our choices, and how to use this to study psychological mechanisms. *Consciousness and Cognition* 86, 103038.
5 Olson, J. A., Amlani, A. A., Raz, A. et al. (2015). Influencing choice without awareness. *Consciousness and Cognition*, Vol. 37: pp. 225–236.
6 Barnhart, A. S. & Goldinger, S. D. (2014). Blinded by magic: eye-movements reveal the misdirection of attention. *Frontiers in Psychology,* Vol. 5 (1461).
7 Fougnie, D. & Marois, R. (2006). Distinct capacity limits for attention and working memory: Evidence from attentive tracking and visual working memory paradigms. *Psychological Science,* Vol. 17 (6): pp. 526–534.
8 Gregoriou, G. G., Gotts, S. J., Zhou, H. et al. (2009). Long-range neural coupling through Synchronization with attention. In: Narayanan Srinivasan (ed.), *Attention Progress in Brain Research*, Vol. 176. Elsevier: Amsterdam.
9 Wisal, K. (2023). The Modern Roman Circus: Control, Distraction, and Manipulation in the 21st Century. *Medium.* https://tinyurl.com/49dydffa
10 Kemp, S. (2024). The time we spend on social media, DataReportal, 31 January.
11 Woolley, K. & Sharif M. A. (2000) Psychology of Your Scrolling Addiction, *Harvard Business Review*
12 Chu J., Qaisar S., Shah Z. et al. (2021). Attention or Distraction? The Impact of Mobile Phone on Users' Psychological Well-Being. *Frontiers in Psychology*, vol. 12: article 612127.

13 Prensky, M. (2001) Digital Natives, Digital Immigrants, Part 2: Do They Really Think Differently. *On the Horizon*, Vol. 9 (6): pp. 1–6.
14 Knoll, J., Matthes, J. & Heiss, R. (2020). The social media political participation model: A goal systems theory perspective. *Convergence: The International Journal of Research Into New Media Technologies* 26 (1): pp. 135–156.
15 Matthes, J., Heiss, R. & van Scharrel, H. (2023). The distraction effect. Political and entertainment-oriented content on social media, political participation, interest, and knowledge. *Computers in Human Behaviour*, vol. 142.
16 Helsper, E. J., & Reisdorf, B. C. (2017). The emergence of a 'digital underclass' in Great Britain and Sweden: Changing reasons for digital exclusion. *New Media & Society,* Vol. 19 (8): pp. 1253–1270.
17 Skovsgaard, M. & Andersen, K. (2020). Conceptualising news avoidance: Towards a shared understanding of different causes and potential solutions. *Journalism Studies* Vol. 21 (4): pp. 459–476.
18 Sidoti, O., Dawson, W., Gelles-Watnick, R., et al. (2024) Social Media Fact Sheet, Pew Research Centre. November. Although this survey involved only US digital immigrants, there is no reason to suppose similar percentages would not be found in other parts of the world.
19 Skovsgaard, M., & Andersen, K. (2020). Conceptualising news avoidance: Towards a shared understanding of different causes and potential solutions. *Journalism Studies*, Vol. 21 (4): pp. 459–476.
20 Atchley, P. (2010). You Can't Multitask, So Stop Trying. *Harvard Business Review*, 21 December.
21 Loh, K. K. & Kanai, R. (2014). Higher Media Multi-Tasking Activity Is Associated with Smaller Gray-Matter Density in the Anterior Cingulate Cortex. *PLOS One* 9 (9).
22 Skovsgaard, M., & Andersen, K. (2020). Conceptualising news avoidance: Towards a shared understanding of different causes and potential solutions. *Journalism Studies*, Vol. 21 (4): pp. 459–476.
23 Personal communication.
24 Vosoughi, S., Roy, D. & Aral, S. (2018). The spread of true and false news online. *Science* 359 (6380): pp. 1146–1151.

NOTES

25 Maertens, R. et al. (2023). The Misinformation Susceptibility Test (MIST): A psychometrically validated measure of news veracity discernment. *Behaviour Research Methods*.
26 Domonoske, C. (2016). Students Have 'Dismaying' Inability To Tell Fake News From Real, Study Finds. *NPR*, 23 November.
27 Maertens, R. et al. (2023) The Misinformation Susceptibility Test (MIST): A psychometrically validated measure of news veracity discernment. *Behaviour Research Methods*.
28 Van der Linden, S. (2023). *Foolproof: Why We Fall for Misinformation and How to Build Immunity*. HarperCollins Publishers: pp. 15–16.
29 Vosoughi, S. et al. (2018) The spread of true and false news online. *Science* 359 (6380): pp. 1146–1151.
30 Mosleh, M., Eckles, D. Martel, C. et al. (2021). Perverse Downstream Consequences of Debunking: Being Corrected by Another User for Posting False Political News Increases Subsequent Sharing of Low Quality, Partisan, and Toxic Content in a Twitter Field Experiment. *CHI '21: Proceedings of the 2021 Conference on Human Factors in Computing Systems*.
31 Bad News (2024) Unsigned article in Science and Technology section of *The Economist*, 4 May: pp. 66 -70.
32 Prensky, M. (2001) Digital Natives, Digital Immigrants, Part 2: Do They Really Think Differently. *On the Horizon* 9 (6): pp. 1–6.
33 Loftus, E. F. (2023) Planting misinformation in the human mind: A 30-year investigation of the malleability of memory. *Learning and Memory*: pp. 361–366.
34 Ost, J., Foster, S., Costall, A. et al. (2005). False reports in appropriate interviews. *Memory* 13 (7): pp. 700–710.
35 Broome, F. (2022). The Mandela Effect is NOT False Memories. fionabroome.com/mandela-effect-false-memories/
36 Prasad, D. & Bainbridge, W. A. (2022). The Visual Mandela Effect as Evidence for Shared and Specific False Memories Across People. *Psychological Science* 33 (12): pp. 1971–1988.
37 Braun, K. A., Ellis, R., & Loftus, E. F. (2002). Make my memory: How advertising can change our memories of the past. *Psychological Marketing*, vol. 19: pp. 1–23.
38 Okado, Y. & Stark, C. E. L. (2005). Neural activity during encoding predicts false memories created by misinformation. *Learning Memory*, vol. 12: pp. 3–11.

39 Curran, T., Schacter, D. L., Johnson, M. K., et al. (2001). Brain potentials reflect behavioural differences in true and false recognition. *Journal of Cognitive Neuroscience*, vol. 13: pp. 201–216.

40 McNally, R. J., Lasko, N. B., Clancy, S. A., et al. (2004). Psychophysiological responding during script-driven imagery in people reporting abduction by space aliens. *Psychological Science*, vol. 15: pp. 493–497.

41 Schmolck, H., Buffalo, E. A. & Squire, L. R. (2000). Memory distortions develop over time: Recollections of the O. J. Simpson trial verdict after 15 and 32 months. *Psychological Science*, vol. 11: pp. 39–45.

42 Loftus, E. (2023) Planting misinformation in the human mind: A 30-year investigation of the malleability of memory. *Learning and Memory*: pp. 361–366.

43 Sidis, B. (1898). *The Psychology of Suggestion*. New York: D. Appleton.

44 Lachapelle, S. (2008). From the stage to the laboratory: Magicians, psychologists, and the science of illusion. *Journal of the History of the Behavioral Sciences* 44 (4), pp. 319–334.

45 Banachek, (2000) *Psychological Subtleties* 2nd edition (Houston, TX: Magic Inspirations) and Brown. D. (2000) *Pure Effect: Direct Mindreading and Magical Artistry* 3rd edition (Humble, TX: H&R Magic Books).

46 Olson, J., Amlani, A. & Rensick, S. (2012) Perceptual and cognitive characteristics of common playing cards, *Perception* online publication.

47 American mentalist the late Robert Cassidy (he died in 2017) used this technique, which he named the 'Car and Tree force'. Magician Max Maven has used animal, vegetable or mineral as the categories from which subjects had to choose. Other ones that have been used include man-made vs organic. The most extensive use of this principle is probably in *Sensory Projection* by Colin Miller and Jamie Badman. See also *The Artful Mentalism of Bob Cassidy Vol. 2*; *Orion* by Phedon Bilek and *Psychological Subtleties* by Banachek.

48 Pailhès, A. & Kuhn, G. (2020). Influencing choices with conversational primes: How a magic trick unconsciously influences card choices. *PNAS* 117 (30): pp. 17675–17679.

49 Pailhès, A. & Kuhn, G. (2020) Influencing choices with conversational primes: How a magic trick unconsciously influences card choices, *PNSA* 117 (30): pp. 17675–17679.

50 Brown, D. (2002). *Pure Effect*. H & R Magic Books.
51 When psychologist Aaisha Hafeez, from the Illinois Institute of Technology, replicated Pailhès and Kuhn's study, he reported a similar, if much-reduced, result, with only one out of sixteen (6 per cent) primed participants choosing the three of diamonds. See Hafeez, A. (2023). Replication Study: Using Subtle Priming to Influence Card Choices in a Magic Trick. *Illinois Tech Undergraduate Research Journal*, vol. 1: pp. 19-28.
52 Nolsoe, E. (2020) Who buys into celebrity endorsements? YouGov Poll, September.
53 Elberse, A., & Verleun, J. (2012). The economic value of celebrity endorsements. *Journal of advertising Research*, Vol. 52 (2): pp. 149-165.
54 Priester, R. J. & Petty, E. R. (2003). The influence of spokesperson trustworthiness on message elaboration, attitude strength, and advertising effectiveness. *Journal of Consumer Psychology*, vol. 13: pp. 408–421.
55 Pringle, H. (2004). *Celebrity Sells*. Chichester: John Wiley & Sons.
56 Klucharev, V., Smidts, A. & Fernandez, G. (2008). Brain mechanisms of persuasion: how 'expert power' modulates memory and attitudes. *SCAN* 3: pp. 353–366.
57 Berger, J. & Fitzsimons, G. (2008). Dogs on the street, Pumas on your feet: how cues in the environment influence product evaluation and choice. *Journal of Marketing Research* 45 (1): pp. 1–14.
58 Lucas, M. (1999). Context Effects in Lexical Access: A Meta-Analysis. *Memory and Cognition* 27 (3): pp. 375–398.

Chapter 12: Banishing Your Blind Spots

1 Government of India Ministry of Railways (2012). Report of the High-Level Safety Review Committee.
2 Leibowitz, H. W. (1985). Grade Crossing Accidents and Human Factors Engineering. *American Scientists*, vol. 95: pp. 558–562.
3 Verma, R. K. (2013), Trespass Prevention on Mumbai Suburban Railway. Paper presented at the Vancouver Railway Safety Conference.
4 Sidiqque, I. (2018). Way to Reduce Track Deaths. *Mumbai Mirror*.

5 Chabris, C. & Simons, D. (2010). *The Invisible Gorilla: And other ways our intuitions deceive us.* HarperCollins: p. 38.

6 Bossetta, M. (2023). Social Media Digital Architectures: A Platform-First Approach to Political Communication and Participation. In S. Coleman and L. Sorensen (Eds), *Handbook of Digital Politics* (2nd Edition): pp. 226-241.

7 Gil de Zúñiga, H., Week, B. & Ardèvol-Abreu, A. (2017). Effects of the news-finds-me perception in communication: Social media use implications for news seeking and learning about politics. *Journal of Computer-Mediated Communication*, Vol. 22(3): pp. 105-123.

8 Sharma, P. (2024). The Illusion of Value: Deciphering Perception Over Reality in Consumer Behaviour. LinkedIn, 24 February.

9 Kantrowitz, K. (2024). Beware of the fake podcast scam. *CMSWire*.

10 Tuft., G. (2024). Scammer Uses Travis and Jason Kelce's New Heights Podcast to Hijack Gabbi Tuft's Press Release Facebook Accounts. GlobeNewswire, 6 March.

11 Vosoughi., S., Roy, D. & Aral, S. (2018) The spread of true and false news online. *Science,* Vol. 359 (6380): pp. 1146–51.

12 Fan, R., Xu, K.E. & Zhao, J. (2016) Higher contagion and weaker ties mean anger spreads faster than joy in social media. The 'incentive structures and social cues of algorithm-driven social media sites' amplify the anger of users over time until they 'arrive at hate speech'. Preprint at http://arxiv.org/abs/1608.03656 and Fisher. M. & Taub, A. (2018) How everyday social media users become real-world extremists. *New York Times*, 10 October.

13 Lewis, P. (2017). 'Our minds can be hijacked': the tech insiders who fear a smartphone dystopia. *The Guardian*, 6 October, 2017.

14 See: Vincent, J. (2017). Former Facebook Exec says social media is ripping apart society. *The Verge*, 11 December; Stevenson, A. (2018). Facebook admits it was used to incite violence in Myanmar. *New York Times*, 6 November.

15 Kaye, D. (2019). Governments and internet companies fail to meet challenges of online hate—UN Expert. OHCHR. 9 October.

16 Mitler, M. M., Carskadon, M. A., Czeisler, C. A., et al. (1988). Catastrophes, sleep, and public policy: consensus report. *Sleep* 11 (1): pp. 100–109.

17 Mitler, M. M., Carskadon, M. A., Czeisler, C. A., et al. (1988). Catastrophes, sleep, and public policy: consensus report. *Sleep* 11 (1): pp. 100–109.
18 US Nuclear Regulatory Commission (1979). Investigation into the March 28, 1979, Three Mile Island Accident by the Office of Inspection and Enforcement (Investment Report No. 50-320/j79-10). July 1979, NTIS NUREG-0600.
19 Hughes, V. (2016). Is There a Relationship Between Night Shift and Errors? What Nurse Leaders Need to Know, *Athens Journal of Health* 3 (3): pp. 217–228.
20 Rogers, A. E. (2003). Hospital staff nurses regularly report fighting to stay awake on duty. *Sleep*, vol. 26, A424–A425.
21 National Highway Traffic Safety Administration 2024 report.
22 Chaput, J. P. (2014). Personal communication, 20 August.
23 Lewis, D. & Leitch, M. (2015). *Fat Planet – The Obesity Trap and How to Avoid it.* Random House: London.
24 Lewis, D. & Smith, D. (2017) Sleep Deprivation, Research Conducted by Mindlab for a commercial client.
25 Sinclair, A. H., Wang, Y. C. & Adcock, R. A. (2024). First Impressions or Good Endings? Preferences Depend on When You Ask. *Journal of Experimental Psychology: General*, 9 September.
26 Sinclair, A. H., Wang, Y. C. & Adcock, R. A. (2024). First Impressions or Good Endings? Preferences Depend on When You Ask. *Journal of Experimental Psychology: General*, 9 September.
27 LeWine, H. (2024). Does exercising at night affect sleep? *Harvard Health Publishing*, 24 July.
28 Drake, C., Roehrs, T., Shambroom., J. et al. (2013). Caffeine effects on sleep taken 0, 3, or 6 hours before going to bed. *Journal Clinical Sleep Medicine* 9 (11): pp. 1195–1200.
29 Silvani, M. I., Werder, R. and Perret, C. (2022). The influence of blue light on sleep, performance and wellbeing in young adults: A systematic review. *Frontiers in Physiology*, vol. 13: 943108.
30 Attwell., D. & Laughlin, S. B. (2001). An energy budget for signalling in the grey matter of the brain. *Journal of Cerebral Blood Flow Metabolism* 21 (10): pp. 1133–1145; Li, T., Zheng, Y., Wang, Z., et al. (2022). Brain information processing capacity modelling. *Scientific Reports* 12 (1): pp. 1–16.

31 Edelman, M. J. (1964). *The Symbolic Uses of Politics.* Chicago: University of Illinois Press.
32 Robin, C. (2004). *Fear. The History of a Political Idea,* Oxford University Press: p. 230.
33 Parisi, D. R., Sartorio, A. G., Colonnelo, J. R., et al. (2019). Pedestrian dynamics at the running of the bulls evidence an inaccessible region in the fundamental diagram. *Proceedings of the National Academy of Sciences* 118 (50).
34 Hawkins, D. R. (2002). *Power vs Force: The Hidden Determinants of Human Behaviour.* California: Hay House Inc: p. 80.
35 Gable, P. A., Poole, D. & Harmon-Jones, E. (2015). Anger Perceptually and Conceptually Narrows Cognitive Scope, *Journal of Personality and Social Psychology* 109 (1): pp. 163–174.
36 Sorella, S., Grecucci, A., Piretti, L., et al. (2021). Do anger perception and the experience of anger share common neural mechanisms? Coordinate-based meta-analytic evidence of similar and different mechanisms from functional neuroimaging studies. *NeuroImage*, vol. 230: pp. 1–45.
37 Kanaya, Y. & Kawai, N. (2024). Anger is eliminated with the disposal of a paper written because of provocation, *Nature. com Scientific Report* Vol. 14.
38 Matias, J., Belletier, C., Izaute, M., et al. (2022). The role of perceptual and cognitive load on inattentional blindness: A systematic review and three meta-analyses. *Quarterly Journal of Experimental Psychology* (Hove) 75 (10): pp. 1844–1875.
39 Most, S. B., Simons, D. J., Scholl, B. J., et al. (2001). How not to be seen: The contribution of similarity and selective ignoring to sustained inattentional blindness. *Psychological Science* 12 (1): pp. 9–17.
40 Wolfe, J. M. (1994). Guided search 2.0: A revised model of visual search. *Psychonomic Bulletin & Review* 1 (2): pp. 202–223.

Index

accidents and disasters
  aviation 20, 32–33, 49–50, 100, 129–131, 132–133
  balance-driven blind spots 129–131, 132–133
  Challenger Space Shuttle 258–259
  change blindness 49–50
  fatigue-induced blind spots 258–259
  illusional blindness 100
  inattentional blindness 19–20, 32–34
  nuclear 259
  road 19, 32, 259
  at sea 32
  spatial disorientation blind spots 129–131, 132–133
Adcock, Alison 260
Afghanistan 83
African Americans, stereotype bias 76–78
age, and change blindness 46
Alario, Margarita 226
Alcamenes 95
Alexander, Robert 114
Allport, Gordon 75, 207
Alpert, Geoffrey 77
Amazon 186
  AI recruitment tool 218
  prediction trick 212–214
Amlani, Alym 244
amodal illusions 90–93

amodal perceptual completion 91–92
amygdala 84
anamorphic illusions 94–96
anchoring effect 253, 254
Andersen, Kim 233, 234
Anderson, Barton 97–98
Anderson–Winawer lightness illusion 97–98
anger-induced blind spots 257, 267–268
anterior cingulate cortex (ACC) 83–84, 233–234
Anthropogene 222
Archilochus 206
Arcimboldo, Giuseppe 162–163
Aristotle 24, 116
arm-bend blind spots 135–136
Armstrong, Lance 248
Artificial Intelligence (AI) 218–219
Asch, Solomon 61–62
Atchley, Paul 233
attention *see* inattentional blindness
autokinetic illusions 99–100
availability heuristic 254
aviation industry
  accidents and disasters 20, 32–33, 49–50, 100, 129–131, 132–133

airport security 38, 48
balance-driven blind
   spots 129–131, 132–133
change blindness 48–50
expectation blindness 80–81
flight attendants 133
head-up displays
   (HUDs) 49
illusional blindness 100
inattentional blindness 20,
   32–33
landing at wrong air-
   port 80–81
'the leans' 131
maintenance engineer er-
   rors 20
sopite syndrome prob-
   lems 133
spatial disorientation blind
   spots 129–131, 132–133

babies
   face detection 160–161, 164
   microbiome 126–127
   nonverbal communica-
      tion 164–166
bacteria 123–129
Bag Man 198–199
Bahrani, Bahador 198
Bainbridge, Wilma 239–240
balance-driven blind
   spots 129–133
Baldwin, Mark 169
bandwagon effect 254
banishing blind
   spots 251–252
   anger-induced blind
      spots 268

   in every-day life 270–271
   expectation blind spots 79–
      80, 82
   fear-induced blind spots 266
   illusioneering scams 256–257
   magical thinking 269–270
Bankman-Fried,
   Samuel 220–222
Banks, Joseph 28–29
Barnhart, Anthony 228–229
Bastiaanssen, Thomaz 125
Bateman's principle 177
bats 157
Bayes, Thomas 73
Bayesian inference 73–74, 89
Bayliss, William 124
Beaumont, William 123–124
Beck, Adolf 55–57, 172
Becker, Robert 25
Becklen, Robert 25–26
Beebe, Rod 33
Berger, Jonah 196–197,
   248–249
Bernard, Claude 137
Biden, Joe 83
big data analytics 218, 219
binding problem 66
Binet, Alfred 9, 21, 243, 244
binocular vision 155
bladder-driven blind
   spots 133–135
Blom, Tessel 155–156
bodily-driven blind
   spots 123–137
   arm-bend blind spots 135–
      136
   balance-driven blind
      spots 129–133

INDEX

bladder-driven blind spots 133–135
microbiome–gut–brain axis 123–129
power of 136–137
Boring, Edwin Garrigues 102
Bornstein, Brian 70–71
Borum, Randy 206
Bosch, Hieronymus 228
Bossetta, Michael 252
Bostrom, Nick 139
bottom-up attention 30
Bouquet, Carole 39
Bowens, Laticia 50–51
Boyle, Willard 203
brain
  amygdala 84
  anterior cingulate cortex (ACC) 83–84, 233–234
  binding problem 66
  brain–gut axis 123–129
  categorisation 45
  deep brain stimulation 169–170
  energy consumption 31–32
  eye movements 145–146
  facial recognition 112, 169–170, 171
  false memories 241–242
  fast-moving objects 155–156
  functional localisers 170
  fusiform face area 169–170, 171
  fusiform gyrus 112
  homunculus concept 149
  inattentional blindness 23–32
  inference 73–74, 89, 103–104
  medial temporal lobe 241–242
  multitasking 233–234
  neural fatigue 42
  pattern recognition 269
  posterior orbitofrontal cortex 83–84
  prefrontal cortex 242
  prosopagnosia 171
  response to failure of positive expectations 83–84
  right inferior parietal lobe 149
  signals from eyes 150–153
  subliminal images 198
  thalamus 84
  'wagon wheel' illusion 147–149
brainstorming 210
breast milk 127
Briscoe, Robert 90
Broome, Fiona 239
Brown, Derren 245
Brown, Nicole 242
Browning, John 186
Bruce, Vicky 159
Bruin, Leon de 65
Buñuel, Luis 39
Burton, Mike 163
business blind spots 203–224
  computers 214–219
  fox/hedgehog concept 206–207
  functional fixedness 209–210
  givens 208–210, 212
  goals 211–212
  Kodak example 203–204, 207, 214

marketing mistake examples 205
operations 210–211, 212
problem-solving blind spots 207–214
Samuel Bankman-Fried scandal 220–222
'butterfly' glioma 113
buying blind spots
arm-bend blind spots 135
choice overload 186–188
familiarity and exposure 198–201
incidental consumer brand encounters (ICBEs) 200–201
online shopping 191
priming 193–198
product placements 195–196
purchasing choices versus decisions 189–190
see also grocery shopping

caffeine 262
Cambridge Analytica scandal 219
Carbon, Claus-Christian 89–90
card tricks
change blindness 34, 43, 52
forcing techniques 226, 244–246
wave change card trick 43
Carter, Travis J. 197
Cartwright-Finch, Ula 24
categorisation 45
CCTV 57

celebrity endorsements 221, 247–248
Cervone, Daniel 25–26
Chabris, Chris 17–18, 26
Chabris, Christopher 251
Chalamet, Timothée 247
Challenger Space Shuttle disaster 258–259
Champ, The (film) 63
Chanel 247
change blindness 35–52
airport security 38, 48
business 207
functional field of view (FFOV) 44–46
Gabor patches 41–42
Hollywood 38–41
magicians 34, 42–43, 52
military 50–51
pilots 49–50
radiologists 38, 46–48
satisfaction of search 38
shopping 201–202
Chaos Monkey program 218
Chaput, Jean-Phillipe 259
charge-coupled devices 203
chemoreception 116
children
microbiome 126–127
teacher expectations 84–85
urination and car journeys 133
see also babies
chocolate 135, 146–147, 187–188, 196
choice blindness 53–66
confabulations 64–65
conformity 61–62

INDEX

consumer goods 60
disgust 62–64
eyewitness identification 54, 55–57, 65
facial preferences 53–54
financial decisions 59–60
moral attitudes/outrage 59, 62–64
taste and smell 57–58
choice overload 186–188, 254
chronophotographic gun 21
Chu, Jianxun 230
Churchill, Winston 206
Clifford, Brian 173
Clinton, Bill 221
Clooney, George 247
Coca-Cola 205
common responses 244
computers
business blind spots 214–219
change blindness study 50–51
impacts on sleep 263
Conan Doyle, Sir Arthur 22
confabulations 64–65
confirmation bias 253, 254
conformity 61–62
Conjurer, The (painting) 228
Conley, Kenny 17–18
consumer goods, choice blindness in 60
continuity editing rules 40
continuity errors 39–41
continuous flash suppression 198
contour completion 92
convictions, wrongful 55–57

copycat brands 202
Correll, Joshua 77
Corstjens, Judith 192
Corstjens, Marcel 192
cortisol 126, 181
court system
eyewitness identification 55–57, 65, 89–90
hungry judges and juries 128–129
wrongful convictions 55–57
Cowan, Nelson 27
Cox, Michael 17
Craver, Nowlin 23
critical thinking 256, 269
cross-modal illusions 116–118
Cryan, John F. 124
cybercrime 191
cybersecurity 219

Danziger, Shai 128–129
Darwin, Charles 62, 175, 206
Dasani water 200, 248
data-quality blind spots 219
decision fatigue 189–190
deep brain stimulation 169–170
deepfake photographs 241
Delboeuf illusions 107–109
Delboeuf, Joseph Remi Leopold 107
dentistry 109, 111
Descartes, René 149
'devil's tuning fork' 104
Diallo, Amadou 76–77
Dickens, Charles 128
Diedrick, Brock 196
digital cameras 203–204, 214

digital natives and digital immigrants 229, 236, 238
Dijksterhuis, Ap 196
disappearing ball trick 67–68
disasters *see* accidents and disasters
disgust, and choice blindness 62–64
distraction techniques 13–14, 227–234
divorce 174
downcodes 217
Drew, Trafton 47
drinks *see* food and drink
Duncker, Karl 209–210
Durlach, Paula 50–51

ear, inner 131–133
Eastern Airlines crash, Miami 50
Ebbinghaus, Hermann 107
Ebbinghaus illusions 107–109
Edkins, Graham 20, 204
Ehime Maru 33
Ejima, Yoshimichi 115
Ekroll, Vebjørn 90, 92
Elberse, Anita 247
elections
　misinformation 235
　priming 197
electromagnetic spectrum 141
'emerging images' 114–116
emotion-induced blind spots
　anger-induced blind spots 257, 267–268
　fear-induced blind spots 257, 264–267

Endeavour 28–29
endolymph 132
Estrin, John 204
E.T. the Extra-Terrestrial (film) 195–196
ethnic stereotype bias 76–78
Evans, Eric 23
evolutionary advantage theory 181
exercise and sleep 262
Exner, Sigmund 31
expectation blind spots 66, 67–85
　Bayesian inference 73–74
　business 206–207, 214
　facial recognition 172
　false memories 240
　how expectations influence perceptions 68–79
　implicit associations 75–79
　magicians 67–68
　optimistic expectations 84–85
　overcoming 79–80, 82
　overconfidence 82–83
　response to failure of positive expectations 83–84
　shopping 202
　stereotypes 75–79
eye-tracking studies
　male and female gaze differences 146–147, 175–176
　roller-coaster rides 265
　shopping behaviour 185–186
　visual Mandela effect 240
eyes 140–146
　binocular vision 155

INDEX

colour and attractive-
  ness 182–183
cones 142, 143
development of 141
eye movements 145–146
feature detectors 150–151
fovea 44, 142, 143–144
functional field of view
  (FFOV) 44–46
optic nerves 110
retina 142–146, 149–151
rods 142–143
saccades and fixations 145–146
sclera 182
signals to brain 150–153
useful field of vision
  (UFOV) 144–145
visual spectrum 141
eyewitness identification 54, 55–57, 65, 89–90
Eyre, Jordyn 172

face detection
  babies 160–161, 164
  development of 161–164
  pareidolia 111–112, 162
Facebook, Cambridge
  Analytica scandal 219
facial attractiveness 178–183
  choice blindness in 53–54
  eye colour 182–183
  similarity 178–180
  symmetry 180–182
facial expressions
  genuine versus polite
    smiles 166
  innate 166
  in subliminal images 169
  understanding 164–166
facial recognition 159–160, 183–184
  brain 112, 169–170, 171
  in CCTV images 57
  context 171–172
  emotional arousal 172–173
  eyewitness identification 54, 55–57, 65, 89–90
  prosopagnosia 171
  in subliminal images 169
facial symmetry 180–182
fake news 235–238
false memories 238–243
false solution paradigm 222
fatigue-induced blind
  spots 257–264
fear-induced blind spots 257, 264–267
fear of missing out
  (FOMO) 230
feature search 65–66
Feldman, Jacob 71
Ferraro, Rosellina 200–201, 248
film industry
  change blindness 38–41
  product placements 195–196
financial decisions, choice
  blindness in 59–60
Fitzsimons, Grainne 248–249
flags, as priming cue 197
'flashbulb' memory 173
FlyDubai Flight 981 crash,
  Russia 129–130
Fockert, Jan de 32

food and drink
  choice blindness in   57–58
  drinks and sleep   262–263
  size illusions   108
  see also grocery shopping
forcing techniques   11, 225–250
  distractions   13–14, 227–234
  misinformation   234–243
  priming   11, 193–198, 243–249
Förster, Jens   136
fovea   44, 142, 143–144
fox/hedgehog concept   206–207
framing effect   254
fraud, online   191
Freud, Sigmund   206
Freudenburg, William   226
frogs   149–150
Frontier supercomputer, Oak Ridge National Laboratory   216
Fry, Eleanor   32
Fugère, Madeleine   147
'fun fear'   264–265
functional field of view (FFOV)   44–46
functional fixedness   209–210
fusiform face area   169–170, 171
fusiform gyrus   112

Gabor, Dennis   41
Gabor patches   41–42
Gaillard, Frank   113
garage sale study   259–260
gender differences
  factors holding attention   146–147
  mate selection   175–177
gender-induced blind spots
  bias against information provided by women   208–209
  stereotype bias   78
Gibson, Daniel   78
global precedence hypothesis of image analysis   97
Goetzinger, Charles   199
Goldman, Ron   242
Gombrich, Ernst   101, 104
Google Glass   205
'gorilla' experiments
  image on medical scans   47–48
  'Invisible Gorilla' video   26, 41
Götz, Friedrich   235
Graybiel, Aston   133
Greenwald, Anthony   79
Gregoriou, Georgia   229
Gregory, Richard   88, 100, 104, 110
Griffith, David Wark   38
Grimes, Anthony   199
Grissinger, Matthew   81
grocery shopping   185–193
  abandoning trolleys at checkout   191
  arm-bend blind spots   135
  boredom   189
  choice overload   186–188
  copycat brands   202
  impulse purchases   189–190
  purchasing choices versus decisions   189–190

INDEX

on-shelf placement of products 192–193
shrinkflation 201–202
store knowledge 190
stress 188
Gulf War 100
gut–brain axis 123–129

Haidt, Jonathan 62–63
Hakidashisara festival, Japan 268
Hall, Lars 53–54, 57–59
halo effect 75, 221
Hamas attacks on Israel (2023) 209
Hamilton, William Rowan 24
Hamlet 112
Hamstra, Petra 198, 248
Harrison, John 117
hatred-based violence 257
Hawkins, David 266
Hawkins, Jeff 156
head-up displays (HUDs) 49
health care *see* medical errors
hedgehog/fox concept 206–207
Helmholtz, Hermann von 29, 73
Hermann, Alexander 22
Hershey 196
Hill, Kyle 89
Hill, William Ely 102
Hitchcock, Alfred 39
Hixon, Gregory 78
Hofstadter, Douglas 123
Hollin, Clive 173
Hollywood, and change blindness 38–41

homunculus concept 149
'honest lies' 64–65
Horberg, Elizabeth 63–64
Humboldt, Alexander von 99
Hürter, Tobias 140
hybrid illusions 96–97

identification
  CCTV 57
  eyewitness 54, 55–57, 65, 89–90
identification parades 55–57, 65
illumination illusions 97–99
illusional blindness 85, 87–122
  anamorphic illusions 94–96
  autokinetic illusions 99–100
  cross-modal illusions 116–118
  Ebbinghaus and Delboeuf illusions 107–109
  hybrid illusions 96–97
  illumination illusions 97–99
  inferences 103–104
  insights from illusions 119–120
  Mach band illusions 110–111
  modal and amodal illusions 90–93
  optic flow problem 106–107
  optic nerves 110
  pareidolia 111–116
  Ponzo illusion 105–106
  shopping 201–202
  size illusions 104–109

trompe l'oeil pictures 101–103
illusioneering 252–257, 264, 266
Implicit Association Test (IAT) 79
implicit associations 75–79
inattentional blindness 17–34, 157–158
  accidents and disasters due to 19–20, 32–34
  brain 23–32
  functional fixedness 209–210
  magicians 20–23, 229
incidental consumer brand encounters (ICBEs) 200–201
Indian Railways 251
infants *see* babies
inference 73–74, 89, 103–104
Inhoff, Albrecht 145
Innocence Project 55
Institute for Safe Medication 81
internet *see* social media
IQ tests 84–85, 260
Israeli intelligence 209
Iyengar, Sheena 187–188

Jacobson, Lenore 84–85
jam-choice studies 57–58, 187
James, William 71
Jastrow, Joseph 21–22
Jazayeri, Mehrdad 68–69
Jenness, Arthur 61
Jin, Linfeng 81
Johansson, Petter 53–54, 58–59
judges and juries *see* court system

Kanal, Ryota 233–234
Kanaya, Yuta 268
Kanizsa, Gaetano 92
Kanizsa triangle 92–93, 110
Kanwisher, Nancy 170
Kardashian, Kim 247
Kawai, Nobuyuki 268
Keil, Matthias 159
Kellar, Harry 22
Kemp, Simon 229
Kennedy, John F., Jr. 130–131
Kepler, Johannes 149
keyhole surgery 109
Kinsey, Michael 33
Klucharev, Vasily 248
Knepton, James 133
Kobylkov, Dmitry 160–161
Kodak 203–204, 207, 214
Kolers, Paul 25
Krieger, Linda 79
Kuhn, Gustav 68, 245–246
Kuleshov, Lev 39
Külpe, Oswald 88

Laeng, Bruno 183
Laing, R. D. 64
Lanford, Laura 116
laparoscopy 109
Larson, Jeffrey 192
Laurence, Sarah 172
Lavie, Nilli 32

INDEX

legal system
  eyewitness identification 55–57, 65, 89–90
  hungry judges and juries 128–129
  wrongful convictions 55–57
  see also police
Leonardo da Vinci 143–144
Lepper, Mark 187
Lettvin, Jerome 'Jerry' 149–151
Levin, Daniel 35, 38, 40–41
Livingstone, Margaret 144
Lo, Edmund 81
Loftus, Elizabeth 238, 242–243
Loh, Kep Kee 233–234
long-term memory 26
loss aversion 205, 253, 254
love at first sight 173–174, 178

Maasalo, Ida 134
McAuliff, Bradley 70–71
McBroom, Buddy 33
McCartney, Paul 171–172
MacDonald, John 118
McGurk, Harry 118
Mach band illusions 110–111
Mach, Ernst 110
McLaughlin, Owen 59–60
Maertens, Rakoen 235, 236
magical thinking 269–270
magicians 9–11, 13–15
  Amazon prediction trick 212–214
  change blindness 34, 42–43, 52
  disappearing ball trick 67–68
  distraction techniques 13–14, 227–229
  expectation blind spots 67–68
  false solution paradigm 222
  forcing techniques 225–229, 234–235, 243–246
  misdirection 11, 13–14, 23, 227–229
  misinformation 234–235
  priming 243–246
  pseudo-explanations 67, 225
  psychologists' study of 20–23
  spoon-bend trick 225
  vanishing coin illusion 68
  wave change card trick 43
magnetoception 116–117
majority view, conforming to 61–62
Malmgren, Philippa 201
Malone, Cara 191
Mandela effect 239–241
Marey, Étienne-Jules 21
Mars chocolate bar 196
Martin, Alexis St 123–124
Marx, Karl 206
Maskelyne, Jasper 10
mate selection differences 175–177
Matrix, The (film) 196
Matsumoto, David 166
Matthes, Jörg 231–234
Maurer, Daphne 180
medial temporal lobe 241–242
medical errors
  change blindness 38, 46–48

dentistry 109, 111
expectation blindness 81
fatigue-induced blind
 spots 259
illusional blindness 109, 111,
 113–114
keyhole surgery 109
radiology 38, 46–48, 111,
 113–114
Meissonier, Ottilie 55–56,
 172
melatonin 262, 263
memory
 false memories 238–243
 'flashbulb' 173
 long-term 26
 Mandela effect 239–241
 misinformation 238–243
 shopping 190–191
 short-term 26–27, 42
 visual system 153–155
 working 27, 190, 229
mental 'laziness' 244
mental priming
 force 245–246
mere exposure effect 198–199
Meyer, Wilhelm 57
Michaels, Bret 256
microbiome 123–129
military
 change blindness study 50–
 51
 Second World War use of
 magicians 10
milk oligosaccharides
 (HMOs) 127
mind–body connection *see*
 bodily-driven blind spots

mindfulness techniques 79
misdirection 10, 11, 13–14,
 23, 227–229
misinformation 234–243
Misinformation Susceptibility
 Test (MIST) 235–236
modal illusions 90–93
modal perceptual
 completion 92–93
Molina, Angela 39
Mona Lisa (painting) 143–
 144, 167, 168
Mooney, Craig 167–168
Mooney faces 167–168
moral attitudes/outrage, and
 choice blindness 59,
 62–64
Morceau de Maurpertuis,
 Pierre Louis 139
Morey, Candice 27
Mosleh, Mohsen 237
Mulholland, John 10
multitasking 233–234
'My Wife and My Mother-
 in-Law' cartoon 102–103

Nagel, Thomas 157
NASA Mars landing 196
national flags, as priming
 cue 197
Necker, Louis Albert 103
Neisser, Ulric 18, 25
nerves
 optic 110
 vagus 126
 vestibulocochlear 132
Nespresso 247
Netflix 186, 218

INDEX

neural fatigue 42
Newcomb, Theodore 220
news, fake 235–238
Nicolls, Michael 102–103
Nightingale, Kate 188
Nisbett, Richard 60
Nittono, Hiroshi 60–61
nociception 116
Noë, Alva 139, 149, 151, 153
Nokia 196
normative conformity 61–62
North by Northwest (film) 39
nuclear accidents 259
nudity 88–89

Okado, Yoko 241–242
Oliva, Aude 70
Oliver, Jamie 248
Olson, Jay 227, 244
online hate 257
online shopping 191
optic flow problem 106–107
optic nerves 110
O'Regan, Kevin 151, 153
Osborn, Alex 210
O'Shea, Robert 109
Ost, James 239
Osteryoung, Jerry 207–208
overcoming blind spots *see* banishing blind spots
overconfidence bias 82–83, 240
oxytocin 125

Packard, Vance 194
Pailhès, Alice 227, 245–246
Pamplona, Spain 265–266
Papies, Esther 198, 248

pareidolia 111–116, 162
parental investment theory 177
Parisi, Daniel 265–266
Park, Choong Whan 190
Parrhasios 101
Parton, Dolly 256
Parvizi, Josef 169–170
pattern recognition 269
paying attention *see* inattentional blindness
Penton-Voak, Ian 178–180
perceptual completion 90–93, 110, 164
Perrin, Steven 62
perspective 95
Phelps, Michael 248
Phidias 95
physical attraction 173–183
  eye colour 182–183
  facial symmetry 180–182
  gender differences 175–177
  love at first sight 173–174, 178
  similarity 178–180
Pickwick Papers (Dickens) 128
Pierson, Ed 20
Pietzsch, Martin 82
pilots
  accidents and disasters 32–33, 49–50, 100, 129–131, 132–133
  balance-driven blind spots 129–131, 132–133
  change blindness 49–50
  expectation blindness 80–81
  head-up displays (HUDs) 49

illusional blindness 100
inattentional blindness 32–33
landing at wrong airport 80–81
spatial disorientation blind spots 129–131, 132–133
Pinquart, Martin 82
Plato, Jason 259
Pliny 95
police 18–19
identification parades 55–57, 65
stereotype bias 76–78
political donations 221
political participation and knowledge
priming 197
social media 231–234, 235
Ponzo illusion 105–106
Ponzo, Mario 105
Pope, Alexander 128
position effect 60–61
posterior orbitofrontal cortex 83–84
Prahl, Andrew 215–216
Prasad, Deepasri 239–240
prefrontal cortex 242
Prensky, Marc 231, 238
Price, Richard 73
primacy effect 259–260
priming 11, 243–249
mental priming force 245–246
semantic 197–198
shopping 193–198
subliminal 193–196, 198
Pringle, Hamish 248

Pringle, Heather 45–46
prior beliefs 70
problem-solving blind spots 207–214
product placements 195–196
Project MKULTRA 10
Propp, Kathleen 208–209
proprioception 116
prosopagnosia 171
pseudo-explanations 64–65, 67, 225
Ptak, Radek 28
Puck (magazine) 102
Puma trainers 249

Quintilianus, Marcus Fabius 24

racial stereotype bias 76–78
radiology 38, 46–48, 111, 113–114
Radon, Mark 47, 113
Read, Don 27
reading 145–146
reality 139–140
Umwelt theory 156–158
Red Mist (television programme) 267–268
Reed, Richard 227
Reid, Vincent 160
Reiss, Spencer 186
Rensink, Ronald 42, 68, 244
restaurants 18, 234
retina 142–146, 149–151
cones 142, 143
feature detectors 150–151
fovea 44, 142, 143–144
rods 142–143

Richardson, Barrie 225
Richler, Jennifer 161–162
right inferior parietal lobe 149
right-side preference 60–61
road accidents 19, 32, 259
Robert-Houdin, Jean-Eugène 10
Robin, Corey 264
robot dog 74
Rodgers, Emily 188
Rogers, Ann 259
roller-coaster rides 264–265
Rosenthal, Robert 84–85
Rubio, Angelica 111
'Running of the Bulls' festival, Pamplona, Spain 265–266

Sacks, Oliver 171
Sagana, Anna 54
saliency 45, 47
Sanders, Donnie 76
Sartre, Jean-Paul 65
Sasson, Steve 203–204, 207, 214
satisfaction of search (SOS) 38
scams, social media 255–256
Schall, Jeffrey 65–66
Schanzer, Jessica 60
Schmolck, Heike 242
Schneider, Maria 39
Schwartz, Barry 187
Schweizer, Kaspar 99
sclera 182
Second World War 10
Segatella bacteria 127

self-fulfilling and self-negating prophecies 207
semantic priming 197–198
senses 116–117
serotonin 125
Shapiro, Scott 216
Sharma, Priyanshu 253
Shelton, Blake 256
Shepard, Richard 106
Shepard, Roger 87
Sherif, Marissa 229
shooter bias 77–78
shopping *see* buying blind spots; grocery shopping
short-term memory 26–27, 42
shrinkflation 201–202
Sidis, Boris 243
similarity-attraction effect 220–221
Simons, Daniel 17–18, 26, 35, 38, 40–41, 251
Simpson, O. J. 242
Sinclair, Alyssa 259–260
Sinha, Pawan 164
size illusions 104–109
SKIMS 247
Skovsgaard, Morten 233, 234
sleep
  impacts of deprivation 259–262
  strategies for improving 262–263
smiles
  genuine versus polite 166
  social contagion 239
Smith, George E. 203
Smith, John 56–57
social contagion 239

social media
  commerce 191
  digital natives and digital immigrants 229, 236, 238
  distraction techniques 229–234
  'doom-scrolling' 229
  'the dress' debate 71–72
  'echo chamber' 82, 254
  fake news 235–238
  fear of missing out (FOMO) 230
  illusioneering 252–253, 254, 255–256
  misinformation 235–238
  online hate 257
  political information 231–234, 235
  scams 255–256
  time spent using 229
  trompe l'oeil artwork study 101–102
Somerville, Jason 59–60
sopite syndrome 133
Sorensen, Herb 192
spatial disorientation blind spots 129–133
speed, underestimating 106–107
Spencer, Christopher 62
spoon-bend trick 225
Stafford, Tom 199
Stark, Craig E. L. 241–242
Starling, Ernest 124
stereotypes 75–79
stereotypical behaviour 244
Sternschwanken 99
Stewart, Doug 107

Strathie, Ailsa 172
stretch reception 116
Strijbos, Derek 65
Stuart, Charles Edward 95–96
subliminal priming 193–196, 198
sunk cost fallacy 254
supermarkets *see* grocery shopping
surgery, keyhole 109
'swapped faces' study 53–54
symmetry, facial 180–182
synesthesia 117
synuclein 125

taste and smell, choice blindness in 57–58
tea-tasting study 57–58
teacher expectations 84–85
testosterone 181
Tetlock, Philip 206, 207
thalamus 84
That Obscure Object of Desire (film) 39
Thatcher, Margaret 167
theme park rides 264–265
Thephakaysone, Nikhom 18–19
thermoception 116
Thomson, Donald 172
Thomson, Peter 167
Thorndike, Edward 75
Three Mile Island nuclear accident, Pennsylvania 259
ticks 156–157
top-down attention 30–31

INDEX

Torralba, Antonio   70
Trainspotting (film)   63
Trendelenburg, Ulrich   124
tricks
   Amazon prediction   212–214
   change blindness   34, 43, 52
   disappearing ball   67–68
   forcing techniques   225, 226, 244–246
   spoon-bend   225
   vanishing coin   68
   wave change card trick   43
Triplett, Norman   67–68, 76
trompe l'oeil pictures   101–103
Tropicana   205
Tuft, Gabbi   256
Tuk, Mirjam   134–135, 136
Tuomisto, Terhi   147
Twain, Mark   22
Tzetzes   95

Uexküll, Jakob von   156–157, 158
Umwelt theory   156–158, 270
unconscious inference   73–74, 89, 103–104
United Flight 173 crash, Portland, Oregon   32–33
upcodes   216–217
urination   133–135
useful field of vision (UFOV)   144–145
USS *Greeneville*   33
Usual Suspects, The (film)   40

vagus nerve   126
Valdez, Justin   18–19
van der Linden, Sander   236

van Rullen, Rufin   148–149
van Swol, Lyn M.   215–216
van Tonder, Gert   115
vanishing coin illusion   68
ventriloquists   117–118
Verleun, Jeroen   247
Verma, Rajendra Kumar   251
vestibular apparatus   131–132
vestibulocochlear nerves   132
Vicary, James   194
Vick, Michael   248
visual illusions *see* illusional blindness
visual Mandela effect   239–241
visual system   139–156
   attention   29–31
   binocular vision   155
   cones and rods   142–143
   development of eyes   141
   expectations   69–74
   eye movements   145–146
   factors holding attention   146–147
   fast-moving objects   155–156
   feature detectors   150–151
   fovea   44, 142, 143–144
   functional field of view (FFOV)   44–46
   homunculus concept   149
   illusional blindness   88–93, 98–99, 103–104, 110–111
   inference   73–74, 89, 103–104
   memory   153–155
   optic nerves   110
   predictions   155–156

processing delay 155
reality 139–140
retina 142–146, 149–151
saccades and fixations 145–146
signals to brain 150–153
useful field of vision (UFOV) 144–145
visual spectrum 141
'wagon wheel' illusion 147–149
visualisation technique 269–270
Vosoughi, Soroush 235, 237
Voss, Joel 112
voting
  misinformation 235
  priming 197

Waddle, Scott 33
'wagon wheel' illusion 147–149
Watanabe, Takeo 112
Watson, Thomas 214
wave change card trick 43
Weger, Ulrich 145
Wickelgren, Wayne 208
Wiederman, Michael 177
Williams, Damian 221
Willingham, Bob 166
Wilson, Tim 60
Winawer, Jonathan 97–98
Wineburg, Sam 235–236
Wise, Jerry 132
Witzel, Christoph 72, 158
Wolfe, Jeremy 31, 48, 139
Woods, Tiger 248
Woolley, Kaitlin 229
working memory 27, 190, 229
Wright, Evan 100
wrongful convictions 55–57

X-rays *see* radiology

Yao, Richard 42
YouTube 253

Zajonc, Robert 199
Zelenska, Olena 237
Zeuxis 101
Zsok, Florian 174